Praise for *Math With*

"With charm, unwavering enthusias..., *Math Without Numbers* waltzes the reader through a garden of higher mathematics."

—Jordan Ellenberg, professor of mathematics,
University of Wisconsin–Madison, author of *How Not to Be Wrong*

"So delightful! Mathematics is playful, surprising, and enchanting, but those qualities are often obscured behind intimidating equations and formalism. Milo Beckman brings them out into the open for everyone to share."

—Sean Carroll, author of *Something Deeply Hidden:
Quantum Worlds and the Emergence of Spacetime*

"*Math Without Numbers* explores deep mathematical topics—and shows how mathematicians think—in completely readable prose. The puzzles and games are bonuses. Very enjoyable."

—Will Shortz, crossword editor, *The New York Times*

"The book's accessible language and illustrations make understanding some of the most complex (and possibly most intimidating) math concepts feel as effortless as breathing. Beckman's approachable writing and Erazo's delightful illustration combine to tell an insightful and entertaining story about math."

—Giorgia Lupi and Stefanie Posavec,
coauthors of *Dear Data* and *Observe, Collect, Draw!*

"This is the book for you if you've ever been curious about the wonderful ideas and concepts underlying modern math but have been too frightened to make a start. Milo Beckman gives us a friendly introduction to unfamiliar concepts and ideas that show why modern math is such a fascinating and rewarding branch of human thought."

—Graham Farmelo, author of *The Universe Speaks in Numbers*

Math Without Numbers

MILO BECKMAN
Illustrated by M Erazo

DUTTON

DUTTON

An imprint of Penguin Random House LLC
penguinrandomhouse.com

Previously published as a Dutton hardcover in January 2021

First Dutton trade paperback printing: January 2022

Illustrations by M Erazo

Photograph on page 187 © Michael Tuchband,
University of Colorado Boulder

LIBRARY OF CONGRESS CATALOGING-IN-PUBLICATION DATA
has been applied for.

Dutton trade paperback ISBN: 9781524745561

Printed in the United States of America

BOOK DESIGN BY M Erazo

to Eriq, for making me do this

with thanks to Taylor for checking my math,
Portia for the dialog, and M for bringing it
to life

what do mathematicians believe?

We believe math is interesting, true, and useful (in that order).

We believe in a process called "mathematical proof." We believe the knowledge produced by proof is important and powerful.

Fundamentalist mathematicians believe that everything—plants, love, music, *everything*—can (in theory) be understood in terms of math.

Contents

Math Without Numbers

Topology

shape

manifolds

dimensions

shape

*M*athematicians like to overthink things. It's sort of what we do. We take some concept that everyone understands on a basic level, like symmetry or equality, and pick it apart, trying to find a deeper meaning to it.

Take shape. We all know more or less what a shape is. You look at an object and you can easily tell if it's a circle or a rectangle or whatever else. But a mathematician would ask: What *is* a shape? What makes something the shape it is? When you identify an object by shape, you're ignoring its size, its color, what it's used for, how old it is, how heavy it is, who brought it here, and who's responsible for taking it home when we leave. What are you not ignoring? What is it that you're getting across when you say something is shaped like a circle?

These questions are, of course, pointless. For all practical uses, your intuitive understanding of shape is fine—no significant decision in your life will ever hinge on how exactly we define the word "shape." It's just an interesting thing to think about, if you have some extra time and you want to spend it thinking about shapes.

Let's say you do. Here's a question you might think to ask yourself:

How many shapes are there?

It's a simple enough question, but it isn't easy to answer. A more precise and limited version of this question, called the generalized Poincaré conjecture, has been around for well over a century and we still don't know of anyone who's been able to solve it. Lots of people have tried, and one professional mathematician recently won a million-dollar prize for finishing up a big chunk of the problem. But there are still many categories of shapes left uncounted, so we still don't know, as a global community, how many shapes there are.

Let's try to answer the question. How many shapes are there? For lack of a better idea, it seems like a useful thing to do to just start drawing shapes and see where that takes us.

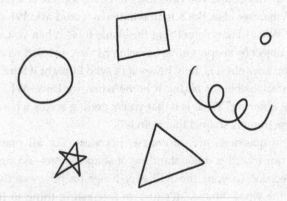

It looks like the answer to our question is going to depend on how exactly we divide things into different shape categories. Is a big circle the same shape as a small circle? Are we counting "squiggle" as one big category, or should we split them up based on the different ways they squiggle? We need a general rule to settle debates like this, so the question of "how many shapes" won't come down to case-by-case judgment calls.

There are several rules we could pick here that would all do a fine job of deciding when two shapes are the same or different. If you're a carpenter or an engineer, you'll want a very strict and precise rule, one that calls two shapes the same only if all their lengths and angles and curves match up perfectly. That rule leads to a kind of math called geometry, where shapes are rigid and exact and you do things like draw perpendicular lines and calculate areas.

We want something a little looser. We're trying to find every possible shape, and we don't have time to sort through thousands of different variations of squiggles. We want a rule that's generous about when to consider two things the same shape, that breaks up the world of shapes into a manageable number of broad categories.

New Rule

Two shapes are the same if you can turn one into the other by stretching and squeezing, without any ripping or gluing.

This rule is the central idea of topology, which is like a looser, trippier version of geometry. In topology, shapes are made out of a thin, endlessly stretchy material that you can twist and pull and manipulate like gum or dough. In topology, the size of a shape doesn't matter.

Also, a square is the same as a rectangle, and a circle is the same as an oval.

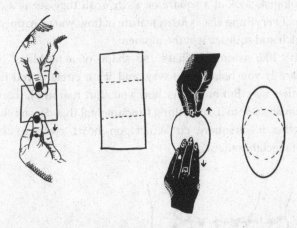

Now it gets weird. If you think about it using this "stretching-and-squeezing" rule, a circle and a square are considered the same shape!

Before you go tell your loved ones that you read a book about math and learned that a square is a circle, keep in mind: Context matters. A square is a circle, *in topology*. A square is most certainly not a circle in art or architecture, or in everyday conversation, or even in geometry, and if you try to ride a bike with square tires you won't get very far.

But right now we're doing topology, and while we're doing topology we don't care about frivolous little details like pointy corners that can be massaged away. We look past superficial differences, things like lengths and angles, straight edges versus curved or squiggly ones. We focus only on the core, underlying *shape*: the basic features that make a shape the shape it is. When topologists look at a square or a circle, all they see is a closed loop. Everything else is just a feature of how you've happened to stretch and squeeze it at the moment.

It's like asking, "What's the shape of a necklace?" It's a square if you hold it one way, and it's a circle if you hold it another way. But no matter how you shift it around, there's an intrinsic *shape* to it, something fundamental that doesn't change, whether it's a square, circle, octagon, heart, crescent, blob, or heptahectahexadecagon.

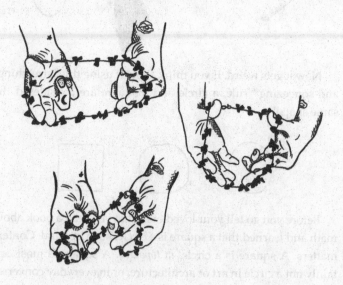

Since this shape comes in many different forms, it's not quite right to call it either a circle or a square. We sometimes call it a circle anyway, but the official name for this shape in topology lore is "S-one." S-one is the shape of a necklace or bracelet or rubber band, a racetrack or circuit, any moat or national border (assuming no Alaskas), the letters O and uppercase D, or any closed loop of any shape. Just like a square is a special type of rectangle, and a circle is a special type of oval, all these shapes are special types of S-one.

Are there any other shapes? It would be a shame if the stretching-and-squeezing rule turned out to be so loose that we accidentally collapsed all the diversity of shapes down into one broad category. Good news: We didn't. There are still shapes that aren't the same as a circle.

Like a line:

A line can be bent almost into a circle, but to finish the job we'd need to click the ends together—not allowed. No matter how you manipulate a line, you'll always have those two special points on either end, where the shape just stops. You can't get rid of end-points. You can move them around and stretch them apart, but the two end-points are an unchanging feature of the shape.

For a similar reason, a figure-eight is a different shape too. There aren't any end-points, but there's still a special point in the middle where the lines cross, where there are four arms reaching out instead of the usual two at any other point. Stretch and squeeze all you want, you can't get rid of a crossing-point either.

If you think about it, this is enough information for us to answer the original "How many shapes are there?" question. The answer is infinity. Here, I'll prove it to you.

Proof

Look at this family of shapes. You make each new shape by adding an extra hatch mark to the previous shape.

Each new shape has more crossing-points and endpoints than all the ones before it. So each one must really be a different, new shape. If you keep doing this forever, you get an infinite family of different shapes, and so there are infinity shapes.

QED

Convinced? All you need to do is find any infinite family of different shapes like this, where it's obvious how to keep making new different shapes forever.

This one would have worked just as well:

Or this one:

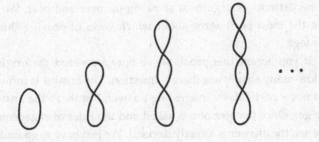

This one works too:

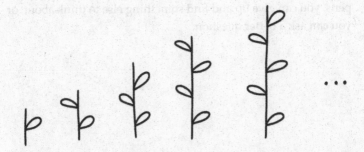

However you prove it, though, the basic argument is the same. You want to show there are infinitely many of something, so you describe a systematic process that keeps churning out new different examples of that thing. This is called an "infinite family" argument, and it's a pretty common tool in math for when you want to show there's infinity of something. I find it convincing—I don't really see how you could argue against it. There have to be infinity of something if you can keep making more of them forever.

And it's not just me: The math community as a whole considers "infinite family" arguments to be valid mathematical proof. There are a bunch of proof techniques like this, where the same sort of argument can be used in different contexts to prove different things. People who do a lot of math start to notice the same patterns of argument showing up over and over. We all (for the most part) agree about which ways of proving things are legit.

If you accept this proof, we've now answered the original "How many shapes are there?" question. The answer is infinity. It's not a particularly interesting answer, but that's the answer we get. Once the question is asked and the rules of engagement are set, the answer is already decided. We just have to go find it.

The first question you think to ask doesn't always lead you to the most interesting or enlightening answer. When that happens, you can give up and find something else to think about, or you can ask a better question.

manifolds

*T*here are too many shapes to keep track of, so topologists focus only on the important ones. *Manifolds*. They sound complex but they're really not—you actually live on a manifold. Circles, lines, planes, spheres: manifolds are the smooth, simple, uniform shapes that seem to always play a leading role when we're working with physical spaces in math and science.

They're so simple, you'd think we would have found them all by now. We haven't. Topologists are so embarrassed about this, they put out a million-dollar bounty to encourage people to look harder. This is the biggest unsolved question in topology, entertaining and frustrating experts in the field for over a century:

How many manifolds are there?

Or, a bit more accurately:

What are all the manifolds?

The goal isn't to literally count them up, but to find them all, name them, and classify them into different species. We're compiling a field guide of all possible manifolds.

So what exactly is a manifold? The rule for qualifying as a manifold is pretty strict, and most shapes don't make the cut.

New Rule

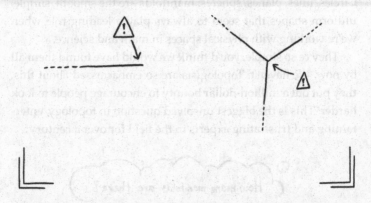

A shape is called a "manifold" if it has no special points: no end-points, no crossing-points, no edge-points, no branching-points. It has to be the same everywhere.

This immediately rules out all those infinite families of shapes from last chapter. Anything with hatch marks or asterisks or anything like that won't count as a manifold. That means the "how many" question might actually have an answer now: There might be an exact, finite number of manifolds. We'll have to see.

This definition also isn't limited to flat, wireframe-style shapes like the ones we've been working with. You can have manifolds made out of sheetlike material, or doughlike material. The universe we live in is probably a three-dimensional manifold, unless you think there's a physical boundary where it just stops, or it crosses over itself somehow.

But let's stick to the wireframe-style shapes for now, the kind you can make out of string or paper clips. In topology we call these shapes one-dimensional, even though the page they sit on is two-dimensional. It's the material of the shape that matters.

So what manifolds can you make out of string? There aren't that many options. Most string-shapes you can come up with have special points.

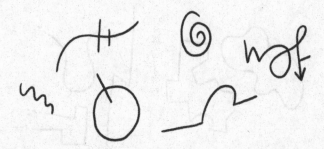

The twists and curls and corners are fine, since those can be smoothed out. The real problem is the end-points. How do you eliminate end-points?

There are only two string-manifolds. If you don't know what they are, you can take a second now to stare off into space and think about it before turning the page.

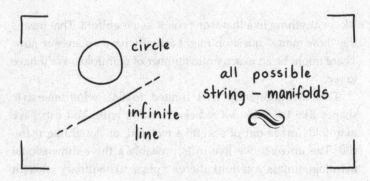

The circle (aka *S*-one) and the infinite line (named *R*-one) are the only manifolds in the first dimension. To avoid end-points, you either have to loop back around or just go on and on forever. And don't forget: Because all the shapes in topology are stretchy, this also covers any closed-loop shape and any goes-on-forever shape. It doesn't have to be literally a circle or a straight line.

That's it for dimension one. Not bad! You can see we've narrowed our search down a lot. The original "how many shapes" question was too big and broad, but this one seems manageable, at least so far. Ready to move up a dimension?

In dimension two, we're looking for manifolds made out of sheetlike material. Remember, it's the material that matters! Most of these shapes are what you'd normally consider three-dimensional, but they're made out of two-dimensional material, and that's where it counts.

So: What manifolds can be made out of sheet-material? We're looking for something that's sheetlike everywhere, with no edges or cliffs where the sheet just stops. Remember how I said you live on a manifold? The surface of the Earth is a sphere, which is a two-dimensional manifold.

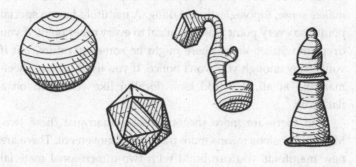

With stretching and squeezing, "sphere" includes any closed surface: cube, cone, cylinder, all the hits. But be careful with your terminology! In math, "sphere" only refers to the hollow surface-shape, whereas a "ball" is filled in. A ball is three-dimensional (made out of dough material), so let's forget about it for now.

This general sphere shape is called *S*-two, which makes sense because it's like the leveled-up version of the circle, *S*-one. We can use the same strategy to find our next sheet-manifold. The equivalent of an infinite line, one dimension up: an infinite plane.

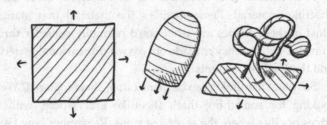

This one's called *R*-two, and it includes any infinite surface that divides space into two infinite regions.

You know how some people think the Earth is flat? That makes sense, topologically speaking. A manifold has no special points, so every point looks identical to every other point, if you drop into Street View. There might be some curvature, but if you're tiny enough you won't notice. If you lived on any sheet-manifold at all, it would look (locally) like you lived on a flat plane.

And there are more sheet-manifolds than just these two. More dimensions means more freedom of movement. There are new manifolds you can build with two-dimensional material that have no equivalent string-shape.

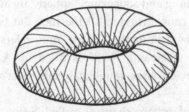

A hollow donut is a manifold. You can tell it's a new manifold because of that hole in the middle—no matter how you stretch and squeeze, you can't get rid of it. But it's a very curious kind of hole: There's no hard edge to it. It's not like you cut a hole out of a piece of paper, leaving a rim of special points. This donut hole is subtler than that. You can only see it from the outside. If you lived on the surface of a donut-shaped planet, you'd never notice from looking around that there was a hole. It would look, locally, just like if you lived on a sphere or a flat plane.

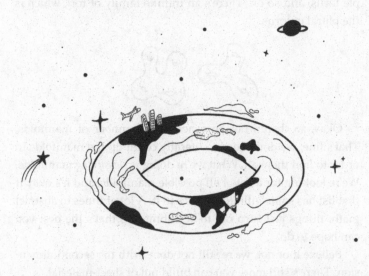

This new manifold is called a torus, or *T*-two, and it includes anything that has this type of smooth hole through it.

We're still not done with sheet-manifolds. You can also make a double torus:

Which of course means you can make a triple torus, quadruple torus, and so on. There's an infinite family of tori, which is the plural of torus.

Okay, so there's not an exact, finite number of manifolds. That's fine, we don't need to literally count up the manifolds in order to find them all. What we're doing is *classifying* manifolds. We're looking for a list of all possible manifolds, and it's okay if that list has some infinite families in it. A lot of times in abstract math, things just turn out to be infinite, so that's the best you can hope to do.

Believe it or not, we're still not done with the second dimension. There's still more you can build out of sheet-material.

There's just a little issue here. The next sheet-manifold I want to tell you about is very strange. I'll tell you what it's called: It's the "real projective plane." But I can't show you what it looks like. I don't know what it looks like. No one knows what it looks like, because it doesn't exist in our universe and never can.

Here's why: It needs a minimum of four dimensions to exist. Regardless of material, each shape has a minimum dimension that it can actually exist in. A plane can fit in two dimensions. A sphere needs three. A "real projective plane" needs four.

So how do we know it exists? Well, let me describe it to you.

Imagine you have a disk, which is a filled-in circle. A disk is made out of sheet-material, but it's not a manifold, because of all the points around the edge. But if you have two disks, you can carefully stitch them together along their edges until they become one shape without any edges at all. They become a manifold.

In this case, that manifold is a sphere, which isn't very helpful since we already know about the sphere. But this basic idea is very useful: You can take two almost-manifolds with the same boundary, and stitch them together to get an actual manifold.

So now imagine you have a thin band of sheet-material with a single twist in it. This shape might look like it has two boundaries, but it only has one, because of the twist. Follow the edge with your finger and you see it loops all the way around the top and bottom and back to where it started.

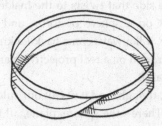

Here's the plan. The boundary of a disk is shaped like *S*-one (a circle). The boundary of this twisted strip is shaped like *S*-one. Let's stitch them together to build a new manifold.

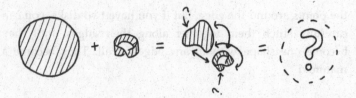

If you try to imagine this in your head or simulate it with your hands, you run into problems pretty quickly. The disk has to twist around and pass through itself, which isn't allowed (no special points). But if you had four dimensions to work in, you'd have no problem.

How's that? Think about a figure-eight. It intersects itself if you draw it on a flat piece of paper, but if you could lift up one of the crossing lines into the third dimension, off the page, it wouldn't intersect itself. Think that, but one dimension up. The weird, twisty manifold we just made intersects itself when we're stuck in three dimensions, but if you could "lift it up" through the fourth dimension, you'd get a perfectly nice, smooth, non-intersecting sheet-manifold.

It's bizarre. This is the real projective plane, or *RP*-two for short, and it's unique and confusing in a couple ways. A sphere and a torus have an inside and an outside, but a real projective plane just has one side that twists to the inside and out. If you write the letter R on a sphere or torus, and slide it around through the space, it'll always come back looking like an R. But if you slide an R around on a real projective plane, it could come back looking like an Я.

It's a manifold, though, and it fits all our rules, so we have to add it to the list. There's the sphere, plane, all the tori, and the real projective plane. Is that it?

Still no. The real projective plane comes with its own infinite family of twisty, unimaginable spaces. Just like you can smush two tori together to get a double torus, you can smush two real projective planes together to get a new manifold called a Klein bottle, which also needs four dimensions to exist without intersecting itself. Or you can smush three of them together, or four, and so you get a whole infinite family of these odd, twisted spaces.

And that, finally, is the complete list of all possible sheet-manifolds.*

Okay, ready to move up another dimension? No, me neither. The next dimension is manifolds made of dough-like material, and even the simplest ones are impossible to imagine. Like the hypersphere, *S*-three, whose *cross-sections* are spheres. So let's not.

You can see how classifying all manifolds could end up being one of the hardest unsolved math problems of all time. The surprising thing is just how little we know. It's not like we made it to dimension ten and got stuck—not even close. Beyond the two dimensions we just looked at, there are question marks all over the place.

The third dimension, dough-type manifolds, is pretty well understood at this point, though it took a hundred years and a million-dollar prize to get there, and we still don't have a totally neat and clean classification like the lower dimensions. In dimensions five and up, topologists use a set of techniques called "surgery theory" to operate on manifolds and construct new ones.

That just leaves dimension four.

I wish I could tell you what's going on in dimension four. I'm not sure there's anyone who really knows. It's a weird boundary case: too many dimensions to do visually, but not enough to use sophisticated surgery tools. There are entire textbooks dedicated to what little we know about four-manifolds, and I couldn't make sense of anything past the opening pages. A professional topologist once told me she'd wanted to work on four-manifolds as an undergraduate but was advised to steer clear.

This is particularly eerie, because many physicists think our universe is best modeled as a four-dimensional manifold, including time as a fourth dimension. If they turn out to be right, that puts some pressure on topologists to get their act together with dimension four. It's not just that we don't know the shape of the universe—until we finish classifying the four-manifolds, the universe might be a shape we haven't thought of yet.

dimensions

When mathematicians talk about the fourth dimension, we're not talking about time. We're talking about a fourth geometric dimension, just like the first three. There's up-down, left-right, forward-back, and then, let's say, "flim-flam." You know, *another one*.

It's pretty clear from looking around, though, that our world has only three spatial dimensions. Don't take my word for it—look at the evidence. If you want to slice a potato into tiny pieces, you need to hold the knife in three different directions.

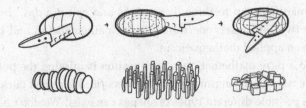

Another way to tell: Imagine you could travel in only two directions. Most of space would be off-limits to you. Any two directions sweep out only a flat plane of motion.

But if you add a third direction, you can travel the whole sky. It takes three directions to cover a three-dimensional space.

One more clue: Consider a jug of any size and shape. If you make a scale replica that's exactly twice as big, it'll hold exactly eight times as much water—double for each dimension.

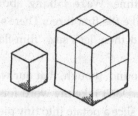

What's the use of talking about an imaginary fourth dimension when we're sure we only have three? Why not just classify the manifolds up to three dimensions and call it a day? I can offer two responses: one from a pure mathematician and one from an applied mathematician.

To a pure mathematician, the question is missing the point. We're not classifying manifolds to be *useful*. We're just curious what possible different types of shapes can exist! We don't have to constrain ourselves to this arbitrary world we happen to live in. Math is general, universal—it's not made in our image. So we have three dimensions. And? We have ten fingers, and do we stop counting there?

That list of sheet-manifolds was out there, somehow, before we ever wrote it down, and it'll still be the complete list of sheet-manifolds long after our civilizations are lost to history. If that alone doesn't make you curious about what types of manifolds exist in higher dimensions, just because they aren't *useful*, well then you weren't really in it for the right reasons to begin with.

Then the applied mathematician comes along and ruins everything by making topology useful.

As it happens, knowing about the topological manifolds is actually useful in quite a few contexts. Yes, even the higher-dimensional ones! It's not why the field was developed, or why people still work on it today, but the language and toolset of topology come in handy pretty often when analyzing aspects of the real world.

The reason it's useful: Humans tend to be visual thinkers, so we often make visual analogies to help us understand abstract ideas. Our everyday languages are so filled with visual analogies that we don't even notice we're using them: You "move forward" on a project, the rents "go up," and an endless argument goes "in circles." When we make these analogies, we're translating real-life problems into topology problems.

Think about politics, for example. Political ideology is a highly complex thing, and it's not always obvious how to compare and contrast two people's beliefs in a concise way. To simplify things, it's common to put ideology on a left-right axis, with progressive, liberal, egalitarian ideals on the left, and traditional, conservative, libertarian views on the right.

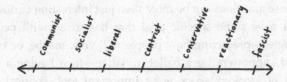

This isn't a perfect system, but it's a useful visual analogy. We can now ask difficult, multifaceted questions in basic, visual terms: "Who's further left on workers' rights?" And sure, we've lost a lot of detail—nothing in the real world is ever as clean and bare as the abstract world of topology—but a lot of what matters is preserved.

Once you've set up a visual analogy like this, you have access to all the language and tools of topology. You can wonder which space is the best representation of the system: Is it a circle or an infinite line? In other words, is ideology cyclical, or can you always move further to the left and right? Or are there special points? Are there "true left" and "true right" positions, and everyone is somewhere in the middle?

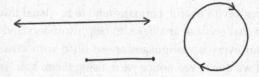

Or maybe we should think there are more dimensions to political ideology than just a single left-right axis. Some people say they're socially liberal and fiscally conservative at the same time. That would mean that ideology space is at least two-dimensional. If that's true, which two-dimensional manifold are we dealing with? Do both axes extend to infinity, like a plane? Does one loop around, making an infinite cylinder? Do they both loop around, like a torus? (Okay, they probably don't both loop around like a torus.)

These questions can be more than just interesting curiosities. If you have some specific goal that has to do with people's ideologies—predicting how people will vote, maybe, or trying to find supporters for a ballot initiative—then having a good model of ideology space is an important tool. Political campaigns use opinion polling to estimate the distribution of voters across an ideology space, and then use these models to tailor their messages and win voters. Political scientists have found a general way to use legislators' voting records to predict how they'll vote in the future, and it works by automatically placing each legislator in a two-dimensional ideology space.

This is how the classification of manifolds is applied outside of math. You only have to solve the abstract math problem once, and then any time you're using a visual analogy to discuss something, the list of spaces to choose from is always the same.

And I really can't stress this enough: We use visual analogies *all the time*. Temperatures rise and fall. Incomes can be low or high or through the roof. December is a long way away, and then it's approaching, and then it flies by, and then it's behind us. All these idioms represent the state of some system as a point in a conceptual space, and then describe changes in that system as physical motion through that space.

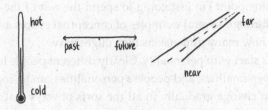

All these examples are one-dimensional, but there can still be interesting topological questions to ask. Does temperature extend forever in both directions, or is there an absolute cold or absolute hot? Does time continue forever to the future, or will there be a Big Crunch? Or does it loop around, so if we wait it out long enough we'll end up in the distant past?

For more complicated concepts, we need to use the higher-dimensional manifolds. Granted, it's pretty rare that we actually need to use the fancier manifolds—the real projective plane, or the triple torus, or the crazy undiscovered manifolds in dimension four. (These come up sometimes in physics, but that's about it, as far as I know.) Most of the systems we come across in our everyday life are well-described by basic, flat spaces: the line, plane, three-space, and so on. In these cases, when we're trying to understand a system, the main topological question is just: "How many dimensions does it have?"

That's the question hiding under the surface in a lot of debates across different areas of discourse. We have some concept. How many dimensions does it have?

When you say that gender is a spectrum rather than a binary, that's a topological claim: You're saying gender space is one-dimensional (a line) rather than zero-dimensional (two separate points). Or maybe you think it's a higher-dimensional space, and the feminine-masculine axis is one axis of difference among many. Questions about which conceptual paradigm to use sometimes boil down to questions of dimensionality.

At this point I'm just going to spend the rest of the chapter going through several examples of conceptual spaces and wondering how many dimensions they might have.

Let's start with personality. Clearly different people have different personalities, and people's personalities can be compared and can change gradually in all the sorts of ways that suggest we might want to use a visual analogy. So what are the dimensions of personality? How can we break down personality into components?

There are lots of personality models to choose from, which come from different intellectual traditions and are used for different purposes and evaluated in different ways. A popular one is the Myers-Briggs personality test, which uses four axes: extroversion–introversion, sensing–intuition, thinking–feeling, and judging–perceiving. Less well-known but preferred by academics is the "Big Five" or OCEAN model, which has five dimensions: openness to experience, conscientiousness, extroversion, agreeableness, and neuroticism. And then there's astrology, which centers on twelve somewhat fluid personality types, each of which manifests in different ways and to different degrees in each person. I suppose you could argue that's something like a twelve-dimensional space.

Which of these models is correct? Well, none of them. Not exactly, at least. Personality is, as far as I can tell, too complicated to be fully described by even as many as twelve dimensions. As with political ideology, we're not hoping to find a perfect description. We just want to get some of the basics down, so we have a common language for talking about and comparing personalities.

Because no model is perfect, each one can be used in different ways by different people for different reasons. For one example, some advertisers use the OCEAN model to design targeted ads on the internet, describing products one way to more conscientious people and another way to less conscientious people. Apparently this model works fairly well for this purpose, but of course if your interest in personality isn't rooted in predicting people's purchasing behavior, you're welcome to use a different model.

It's worth mentioning: All these models have more than three dimensions, and that's not a problem. If you had a good three-dimensional model, you could represent each person as a literal point in a physical three-dimensional space. And sure, you can't do that with four or more, but you can still sort of imagine what that would mean, even if you can't actually picture a twelve-dimensional space.

Here's an example that's much simpler. Let's call it faucet space. What is the space of all possible settings on a standard faucet?

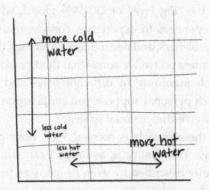

The answer is two. You choose the amount of hot water and the amount of cold water, and that fully describes a faucet setting. For a system like this, the number of dimensions is the same as the number of dials or controls. For this reason, dimensions are sometimes called "degrees of freedom."

But wait: There's another way to slice up faucet space. Some faucets don't have two separate knobs; they have a single handle that moves up–down to control the amount of water, and left–right to control the temperature.

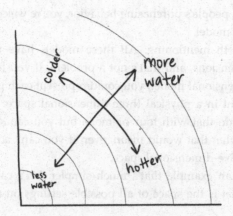

This kind of faucet covers the exact same faucet space as the two-knob kind—they have exactly the same possible water settings. They're just two different ways of doing the same thing. If you want to single out a particular water setting, you could specify the amount of hot water and the amount of cold water, or you could specify the total amount and the temperature. Either way, there are two coordinates. It's a two-dimensional space.

Another household example. I can't understand why my toaster oven has three knobs. As far as I can tell, there are two variables I can control: the temperature, and the amount of time before it goes ding. That would be a two-dimensional space. So why are there three knobs? What is the difference between toast, broil, and bake?

While we're in the kitchen, let's talk about baking. Each recipe specifies an amount of flour, butter, eggs, and so on, and then an oven temperature, and an amount of time. We can think of a full recipe, then, as a point in a high-dimensional space, where each axis corresponds to one ingredient. When you change a recipe by adding more cocoa powder, you're moving the recipe point further out along the cocoa powder axis. When you raise the oven temperature, you get a new recipe further out along the temperature axis.

In this topological model, the vast majority of points represent absolutely disgusting recipes, something like a gallon of baking powder plus one egg. The art of baking can be thought of as a process of testing out different points in this space and trying to find which ones are delicious. There's a region of this baking space that's called "cookies" and a region called "cake" and a smaller region inside there called "pound cake." Of course there are more variables that go into baking than just a list of ingredients—like how soft the butter is when you add it, or how exactly the batter gets arranged in the oven and on what kind of dish—but you can imagine that you might be able to add these as extra dimensions and end up with a pretty comprehensive model of baking as a topological space.

Now maybe you can start to see why some hardcore mathematicians think the entire world is one big math problem. If we can approximate complex concepts fairly well with basic math concepts, who's to say we can't just complicate our models slightly and end up with an exact mathematical description of everything?

Three more quick examples. Taste, we're told, has five dimensions, corresponding to our five types of taste buds: salty, sweet, bitter, sour, and umami. If this is true, then every single flavor you've ever tasted is given by an amount of salty plus an amount of sweet plus et cetera. That feels a little sad and reductive, but on the other hand, it's a good demonstration of just how much room there is in a five-dimensional space.

And besides, it's not quite right to say that a flavor is a single point in that space. When you bite into a taco, you're not tasting a single taste-point. You're experiencing a rapidly changing sequence of different tastes. So maybe it's more accurate to think of each flavor as a *path* through taste space, giving us a lot more room to find new flavors even within the confines of our five basic tastes. After all, our hearing is described by a single variable (pitch, aka frequency) and people keep coming up with new and beautiful ways to pull us around through pitch space over the course of a few minutes.

Color is three-dimensional. You probably learned this as a kid, without phrasing it in terms of dimension. Every color can be made out of three primary colors, combined in different amounts. We figured out that color space was three-dimensional long before we learned why: Our eyes have three different color receptors, each sensitive to a different frequency of light. Your red cones vibrate some amount, your green cones vibrate some amount, and your blue cones vibrate some amount, and that picks out a point in three-dimensional color space, aka a color.

This is why color pickers on computer programs have three dimensions of control. Sometimes they give you three sliders: red, green, and blue. Or sometimes it's hue, saturation, and brightness. Sometimes they give you a two-dimensional disk of colors plus a brightness slider. As with faucet space, there can be multiple ways to choose coordinates, but they all span the exact same color space. And the neat thing about dimensions is that, no matter which coordinate system you pick, each space has a fixed number of dimensions.

I saved the weirdest example for last. As expected, most of these real-world spaces are described well enough by basic, flat spaces, without any closed loops or twists. The weirder manifolds used to be thought of as a sort of intellectual curiosity, something topologists worked on purely for the principle of finding them all. But then people started to realize that the physical universe might be one of these weirder spaces.

Physical space, as we can see, has three dimensions. And time has one dimension. In some areas of physics, it becomes necessary to treat these concepts together as one unified thing, spacetime. Just like you give a friend a time and a place when you're meeting up, physicists identify events in spacetime with four-dimensional coordinates. You might think that spacetime would be the standard four-dimensional space, where each dimension is a straight line. It isn't. At least, when we try to model spacetime as a standard four-dimensional space, it gives us inaccurate predictions.

If space-time is a curved or twisted space, something like the torus or real projective plane, then our intuition about how reality works would break down when we try to consider the universe as a whole. The universe could be finite but have no boundary, like the surface of a sphere. It could expand but not be expanding into anything. There could really be nothing before the Big Bang, just like there's nothing north of the North Pole. Questions about the possibility of time travel, or wormholes that take you immediately from one part of space to another, could come down to which type of space exactly we live in.

Of course, topologists don't care about any of this "applied math" nonsense. They're just trying to find all the shapes.

moon and sun math

If you know which way's east and you know around when the sun rises/sets you can use angles to figure out the time.

A full moon never appears in the day. A new moon never appears at night. Half moons appear half the time in each.

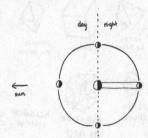

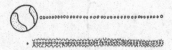

The moon is about a hundred moons away.
The sun is about a hundred suns away.
That's why they're the same size in the sky.

regular polytopes

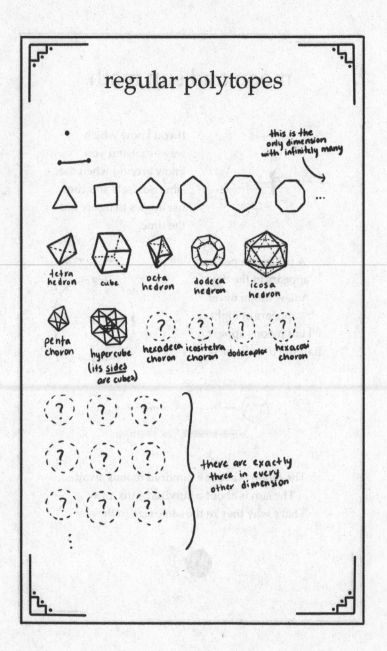

this is the only dimension with infinitely many

tetra hedron cube octa hedron dodeca hedron icosa hedron

penta choron hypercube (its <u>sides</u> are cubes) hexadeca choron icositetra choron dodecaplex hexacosi choron

there are exactly three in every other dimension

38

some facts about circles

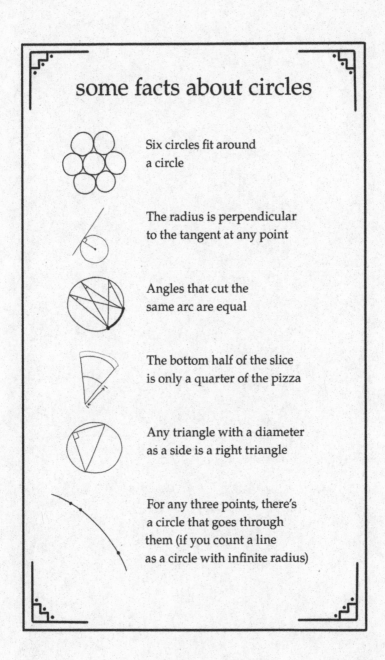

Six circles fit around
a circle

The radius is perpendicular
to the tangent at any point

Angles that cut the
same arc are equal

The bottom half of the slice
is only a quarter of the pizza

Any triangle with a diameter
as a side is a right triangle

For any three points, there's
a circle that goes through
them (if you count a line
as a circle with infinite radius)

Analysis

infinity

the continuum

maps

infinity

𝒴ou know what infinity is. It's bigger than every number. It's what you count toward when you count forever without stopping. It's the entirety of everything that exists and then some.

When people ask about infinity, there's always one thing they want to know:

Is anything bigger than infinity?

This question actually does have an answer. It's not an open question, and it's not a trick question. The answer is either "yes" or "no" and by the end of the chapter I'll tell you which it is.

You can try to guess now, but maybe we should set the rules of the game first so you know what we're talking about.

Specifically, we need a rule for "bigger." How will we know for sure if we've found something bigger than infinity? It's easy to tell with finite quantities when something is bigger than something else. It doesn't seem so obvious with infinity. We don't want to have to settle for judgment calls, so let's pick a solid, foolproof rule for when one quantity is "bigger" than another.

Well, how do we usually settle "bigger," in regular, finite cases? What does it mean to say that the pile on the right is bigger than the one on the left?

Yes, it's totally obvious from looking at it. But imagine you meet someone, some alien from another planet, who's never heard of "bigger," "more," "greater," or anything like that. How do you explain that the right pile is bigger? Really, try it. It's such a basic concept that it's actually hard to spell it out from scratch.

A common trick in math, when you get stuck, is to ask the exact opposite question and see where that takes you. How would you explain to the alien that these two piles are the same size?

You can't lean on the word "equal," since that's exactly what we're trying to define. This alien wants to understand what you're talking about, what the big idea is, when you call things "equal" or "the same."

Here's something you could do to get the point across. Line up the piles and show that they can be paired off one-to-one. They're the same size because you can match them up perfectly, with no leftovers.

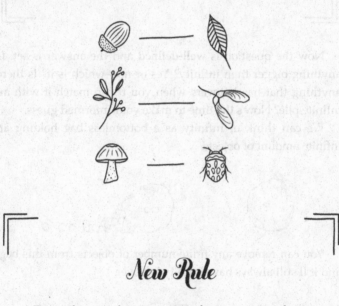

New Rule

Two piles are the same size if you can match up their objects without any leftovers.

The "opposite question" trick worked: We can get a good definition of "bigger" by flipping the rule.

New Rule

If you can't match up two piles perfectly, the side
with leftovers is the "bigger" pile.

Now the question is well-defined and the answer is set. Is
anything bigger than infinity? Yes or no—which is it? Is there
anything that has leftovers when you try to match it with an
infinite pile? Now's the time to make your informed guess.

We can think of infinity as a bottomless bag holding an
infinite amount of objects.

You can remove any finite number of objects from this bag,
and it'll still always have infinity left over.

How could anything be bigger than that? Well, what about infinity plus one?

It doesn't seem that one extra object should make a difference compared to infinity, but let's use the matching rules to make sure. First we can arrange the objects of the infinity bag in a line so it'll be easier to see what's getting matched with what.

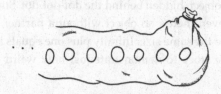

If we try to match things up the obvious way, it certainly looks like infinity plus one is bigger.

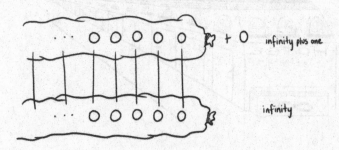

But be careful! Our rules say something is bigger only if you *can't* match them up. (Always good to go back and check the rules.) There's a different way to do the matching that *does* work, without any leftovers on either side:

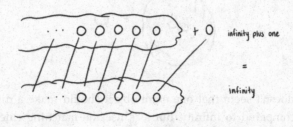

If this seems like cheating, pause to convince yourself it isn't. We're not matching an object with a dot-dot-dot, we're matching it with the next object, hidden behind the dot-dot-dot. Since both bags go on forever, there's no object without a partner, and so the two piles are the same size. Infinity plus one equals infinity!

Let me tell a story to demonstrate just how weird this result is.

Imagine you're the receptionist at a very special hotel, called Hotel Infinity. Hotel Infinity has infinitely many rooms. There's one long hallway, with a line of doors, and the doors go on and on forever, never ending, no matter how far you walk. There's no "room number infinity" or "last room" because there's no end to the hallway. There's a first room, and then for every room, there's a next room over.

Tonight is an especially busy night: Every room in the hotel is full. (Yes, this world has infinity people too.) If you walk down the hallway as long as you like, and knock on a door, you'll hear, "Someone's in here! Do not disturb!" Infinity rooms, filled with infinity people.

Then someone walks into the hotel lobby from the outside world and says, "Could I please have a room?"

It's not your first night at Hotel Infinity, so you know exactly what to do. You get on the PA system and make an announcement: "Apologies for the inconvenience. All guests please move down one room. That's right: Pack up your things, go out into the hallway, and relocate to one room further away from the lobby. Thank you and have a great night." Once everyone does what you say, you've cleared out a room for the new guest.

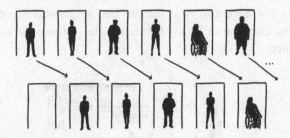

Infinity rooms, infinity plus one guests, and you still have a perfect matching of rooms and guests. Infinity plus one equals infinity.

Infinity plus five, infinity plus a trillion—it doesn't matter. The same logic applies. You can match the bags, you can fit the extra guests. Infinity is so big that finite quantities don't even register in comparison. So we haven't found anything bigger than infinity.

What about infinity plus infinity? Can two infinity bags be matched up with one?

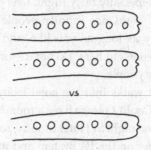

We can't "shift over" this time. We need a new trick if we're going to be able to match these. Or maybe it's impossible to match them, and we've found something bigger than infinity. What do you think?

Here's the same question in Hotel Infinity terms. You're back at the desk, with a full hotel. Into the lobby walks not one new guest but an entire new infinite line of guests, all needing rooms. Can you fit them? Is infinity plus infinity the same as infinity?

Again, the same trick won't work—how can you tell someone to walk down infinity doors? Where would even the first guest end up? There is no "room infinity-plus-one" to relocate to.

Is it possible?

It is possible, and here's how. You get on the PA system again. "Apologies, everyone. Will the guest in the first room please move to the second room, will the guest in the second room please move to the *fourth* room, and in general will everyone walk down to the room that's *twice* as far from the lobby?"

Everyone still has a room, and miraculously, by spacing them out, you've opened up infinity rooms for the new guests. If the doors are numbered, all the odd-numbered rooms are now empty.

Here's the same spacing-out argument in bag-world:

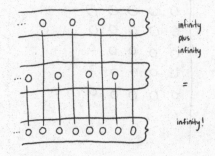

infinity
plus
infinity

=

infinity!

Maybe you think this is a little too much. It's definitely a little counterintuitive, I'll give you that. But if you really want to talk about infinity, you're going to have to question your intuition. You're going to get weird, unintuitive results, like infinity being equal to twice itself. Mathematicians refused for the longest time to work with infinity, because of proofs like this, and plenty of math teachers today will still tell you infinity isn't a number—it's not real math.

But that's the secret about real math: You can study anything at all, as long as you lay out the rules of the game ahead of time. You can work with infinity, if you're clear about what that means and you're willing to swallow some potentially bizarre results. In this case, the rule we chose for "the same" makes it so that infinity plus infinity equals infinity. If you don't like it, I understand, and you're welcome to go back and pick a different rule and re-ask the question. I'm going to stick with the rule we picked.

Infinity plus infinity is infinity. And by the same logic, three infinities or a thousand infinities, it's all still equal to the same, original infinity. Is it time to give up?

Let's try just one more time. Infinity *times* infinity. Is that bigger than infinity?

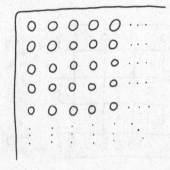

Can you match this up with a single infinity bag?

I'll cut right to the chase this time: You can. It's still the same size. Here's a proof without words.

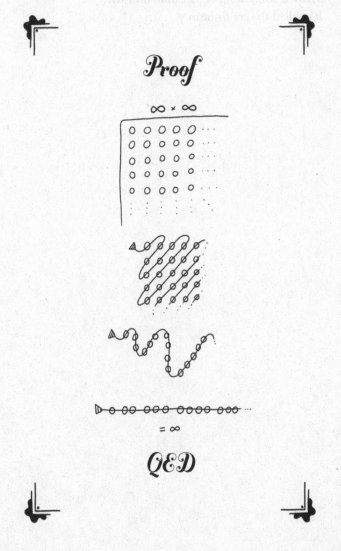

So there you have it: Infinity times infinity equals infinity. We still haven't found anything bigger. So now, as promised, it's time to reveal the answer to the big question.

There *is* something bigger than infinity.

It's called the continuum.

the
continuum

$\mathcal{T}$he continuum is bigger than infinity in the way that infinity is bigger than one. It's unthinkably bigger. It's a different type of bigger. It's so big that regular infinity doesn't even register in comparison.

The continuum is also referred to as "continuous infinity" and it's commonly written as just a lowercase c. Aesthetically, you can think of the continuum as having a smooth, continuous texture, like a ribbon. This contrasts with the infinity from last chapter, which we visualized as a bag of separate objects. That infinity is called "countable infinity" because you can point out each of its individual elements and put them in order.

The continuum is the number of points in a line. It doesn't matter if the line is finite or infinite. It's the texture that matters, the density of points. This is a rich, full, thick type of infinity we're dealing with here. No matter how far you zoom in, it never thins out—a tiny slice of line still has a continuum of points.

It's helpful to compare the continuum to the original, countable infinity, to see just how much bigger it is. Countable infinity is like the whole numbers: a series of dots, evenly spaced out on an infinite line. You can make a two-dimensional grid of points like this, or a three-dimensional lattice, or four or more, and you'll still just have a bunch of separate points. Even if you tighten up the spacing between points, by a factor of a hundred or a million, the points are still separate, and if you zoom in far enough you can pick out a particular one. That's a countable infinity.

The continuum, by contrast, includes all the points in between. *All* of them. It's a vast, smooth sea of points that blend into each other. It's uncountable.

Another way to look at it: If you throw a dart at the number line, the chance of it landing perfectly on a whole number is exactly zero. Not some very small chance—zero. There are infinitely more numbers in the in-between.

This is an important distinction that comes up a lot in math and in the real world: "discrete" versus "continuous." Here are some familiar examples.

discrete   continuous

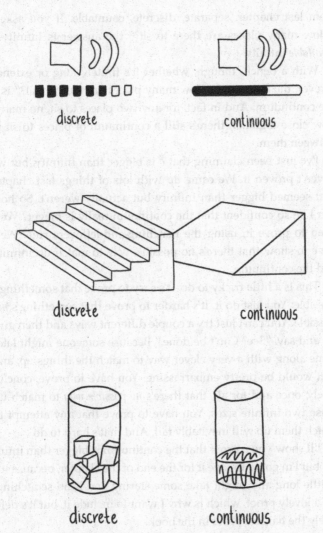

discrete   continuous

discrete   continuous

discrete   continuous

Any discrete collection of things has either finite size or *countably* infinite size. In all these examples it's finite, but imagine you had an endless row of chairs. That's like the bag from last chapter: separate, discrete, countable. If you asked, "How many places are there to sit?" the answer is infinity—*countable* infinity.

With a bench, though, whether it's finitely long or extends forever, the answer to "How many places are there to sit?" is c, the continuum. And in fact, for any two places to sit, no matter how close together, there's still a continuum of places to sit in between them.

I've just been claiming that c is bigger than infinity, but we haven't proven it. We came up with lots of things last chapter that seemed bigger than infinity but actually weren't. So how can I be so confident that the continuum really is bigger? We'd need to prove it, using the matching-and-leftovers rule. We'd have to show that there's no possible way to match up infinity and the continuum.

This is a little tricky to do. It's easy to prove that something's possible: You just do it. It's harder to prove that something's *not* possible. You can't just try a couple different ways and then give up and say, "See? Can't be done." Because someone might later come along with a very clever way to match the things up, and that would be pretty embarrassing. You have to prove, conclusively, once and for all, that there's *no possible way* to match up these two infinite sizes. You have to prove that any attempt to match them up will inevitably fail. And that's hard to do.

I'll show you a proof that the continuum is bigger than infinity, but I'm going to save it for the end of the chapter, because it's a little long and might take some staring and head-scratching. It's a lovely proof, which is why I want to include it, but it's definitely the hardest proof in the book.

Instead, to hold you over, here's another nice proof that's also related to what we're talking about. I told you that the continuum is the same whether it's a finite line or an infinite line. Here's a proof.

Proof

Take a finite continuum and an infinite continuum. Bend the finite one into a semicircle and draw an X at the center. Put the infinite one in a straight line below.

Now here's how we match them up. For any point on the infinite continuum, use a ruler to connect it to the X. This connecting line crosses the finite continuum at exactly one point. Match up that intersection point with the original point on the infinite continuum.

Each point on the infinite continuum matches with exactly one point on the finite one, and vice versa. There are no leftovers on either side, so they're equal.*

QED

You might wonder whether any object as dense and rich as the continuum could actually exist in the real world. There definitely can't be a continuum on a screen, because screens are made of pixels and pixels are discrete, separate objects. In the same way, if our world is made out of tiny particles, there's never really a continuous infinity of anything, except maybe time.

Still, somehow, the continuum is the main character of the most useful area of math outside of basic arithmetic. Most of modern science and economics is built on a single mathematical tool that lets you add together a continuum of numbers and get a finite answer. This tool is called the integral, but I'm going to call it the continuum-sum, because that's what it is.

Here's an idea of how it works. Say you want to measure the length of a curvy path, but all you have is a straight ruler.

You could get a rough estimate of the length by breaking it up into nearly straight segments, measuring each of those, and adding them all together. It wouldn't be exact, but it'd be fairly close.

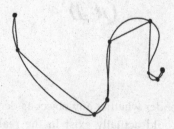

If you need a more precise answer, you can cut the curve up into much smaller pieces, as many as a hundred or even a thousand. Each individual piece will be super tiny, very close to zero, but if you add them carefully and hold on to all the decimal places, you'll get an answer very close to the actual length.

To a mathematician, though, "very close" is still not good enough. We need to know the *exact* length. To get it, we do something that seems like it shouldn't be possible: We cut the curve all the way up into a continuum of pieces, of infinitesimally small point-pieces, and somehow, using the continuum-sum, add them all together.

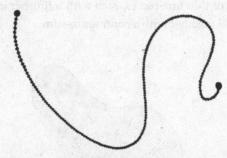

Believe it or not, this is an actual thing we can actually do, and it spits out a finite answer. Not zero or infinity, but an exact length, something like six or pi.

It's a neat trick, and like most mathematical tools, it's general and abstract enough to be applied across many contexts that on the surface have nothing to do with each other. I'll give a couple more examples, but there's really no way I can get across just how versatile the continuum-sum is. It's everywhere.

This is similar to the first example, but let's say you want to find the area of some blob—a pond, say. It's easy to calculate the area of a rectangle, but this isn't a rectangle. You can estimate by slicing it up into thin pieces, each of which is going to end up being pretty close to a rectangle.

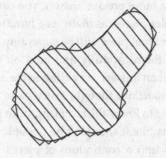

But if you want the exact area, you'll need to slice it into a continuum of thin line-pieces, each with infinitesimal area, and add them all together with a continuum-sum.

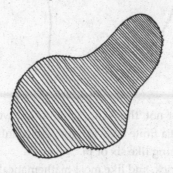

The continuum-sum of a bunch of points makes a line, and the continuum-sum of a bunch of lines makes an area.

Here's an example that looks pretty different, but ultimately the same kind of thing is going on under the hood. Imagine you take an hour-long drive in a car that doesn't track distance—it only has a speedometer. You want to know how far you traveled, based on knowing your speed at each time. Is it possible? How would you do it?

You can get a (very) rough estimate if you look at the speed once during the hour, and assume that's your constant speed for the whole hour. But that's not a very good estimate. What if you started off slow and sped up as the hour went on? You might have checked the speedometer at a moment that's not representative of the whole trip.

You can get a better estimate of your total distance if you split the hour up into shorter periods. You check the speedometer once each period, and that tells you about how far you went in that period. Add up all those distances, and that's about how far you went in the hour.

The estimate gets better and better as you break the hour up into smaller and smaller pieces. Think about it: When you get down to second-long slices, the speed is probably pretty close to constant throughout that second.

See how this is similar to the curve and blob examples? You can get the exact answer by slicing the hour up all the way into a continuum of instants, and add together the speed at each instant with a continuum-sum. A continuum-sum of points is a line, a continuum-sum of lines is an area, and a continuum-sum of speeds is a distance.

You can use this same strategy not just to calculate a distance from speeds, but to calculate any total quantity when all you have is its rate of change. If you want to know the *total* decrease in forest cover, and all you have is the *rate* of deforestation, you can use a continuum-sum.

Of course, if the rate of deforestation (in trees per day) is constant over time, you won't need to be this fancy. You can just multiply the rate by the number of days to get the total. Even if the rate is changing, if you have day-by-day data of how many trees are chopped down, you can still just add them together to get the total. It's only when the rate is changing *continuously*—every split second—that you need the continuum-sum.

That's why the continuum-sum is particularly useful in fields like physics and engineering, where you deal with all sorts of continuously changing quantities: temperatures, water flows, fuel amounts, speeds, electric currents, and so on. But it's such a convenient tool that people have even found ways to use it on discrete quantities like bank accounts, which move in discrete one-cent ticks, or animal populations, which move in discrete one-animal ticks. If you pretend that "wealth" or "population" is a continuous quantity, you can apply all the same predictive techniques physicists and engineers use. You just have to remember to round to a whole number at the end.

And now, since you've waited so patiently, here's a proof that the continuum really is bigger than infinity.

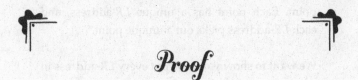

Proof

We're going to prove that any possible attempt to match up the continuum with discrete infinity will fail, leaving leftovers on the continuum side. We'll prove, in other words, that the points of the continuum can't be put into a list, even an infinite list.

We'll use a finite-length continuum, since (remember) the size doesn't matter. Let's give each point a name. The name of each point will be an address that tells you where to find it. The first letter tells you if the point is on the left or right half: *L* or *R*. The second letter tells you if it's on the left or right half *of that half.* And so on, with each *L* and *R* zooming in closer to the point.

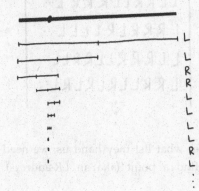

A finite string of *L*s and *R*s only narrows down to a continuous region of the strip. But an infinitely long *LR*-address gets you the location of an exact point. Each point has a unique *LR*-address, and each *LR*-address picks out a unique point.*

We want to show you can't put every *LR*-address in a list, even an infinite list. Imagine your rival comes along with an infinite list and claims that every *LR*-address is on it. We think they're wrong. But we need to prove it.

a supposedly complete
list of all points

LLRRLLLLRL...
LRLRLLRRRR...
LRRRRRRLRR...
LLRRLLRRRL...
LRRRLRLLLL...
LLRRRLLRRLL...
LRRLLRLRLRL...

No matter what list they hand us, we need to be able to find a point (aka an *LR*-address) that's missing.

Here's how we do it. Start at the beginning of their list. Whatever's the *first* letter of the *first* address, write down the opposite. Then, whatever's the *second* letter of the *second* address, write down the opposite. Continue like this down the infinite diagonal.

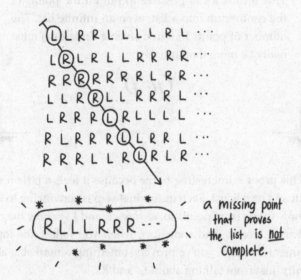

a missing point that proves the list is <u>not</u> complete.

You've now written out a full *LR*-address. We claim this *LR*-address is missing from your rival's list. How do we know? Well, it can't be the first address in the list, since they disagree on the first letter (at least!). It can't be second in the list either, since they disagree on the second letter. It's not the billionth address in the list, since it has the wrong billionth letter.

It can't be anywhere in your rival's list at all.

It doesn't matter what list your rival hands us—we can always use this technique to find a missing point. Even if they take our missing address and insert it at the top, we can just run the process again to find a new one.

This means it's impossible to put all the points of the continuum into a list, even an infinite list. The number of points in a line (even a finite line) must really be more than infinity.

QED

This proof is interesting to me because it feels a little round-about and backward. Each individual step is convincing to me: I see how we go from points to addresses, and I see how the diagonal trick works. And somehow, when you follow the logical argument through, you've proven something remarkable about infinity. Just from talking about *L*s and *R*s.

If you accept this proof, then there really is something bigger than infinity. There's not just finite and infinite, there's another layer on top of that. Which raises *so* many questions. Is there anything between infinity and the continuum, or is the continuum the "next" bigger thing? Is anything bigger than the continuum? How many different infinite sizes are there? Are there a finite amount or infinite amount of infinities? And if it's infinite . . . which type?

Some of these questions have answers and some don't. The first question (whether there's something between infinity and the continuum) has turned out to be the strangest of all. It really seems like a yes-or-no question—there is or there isn't. But someone found the answer, and proved it, and it's neither yes nor no.

Little-known fact: There's a third, more avant-garde status between true and false. But I can't tell you about that yet.

maps

I have to come clean: Most of the content in the last two chapters isn't technically classified as analysis. It's more like a prelude to analysis. Actual analysis deals with infinity and the continuum the way journalists deal with vowels and consonants: They're there, and you have to know what they are and how they work, but that's not really the focus. Analysis is mostly about maps.

A map, in the standard everyday sense of the word, is a picture where dots or symbols are understood to correspond to real-world places and things. They're not just markings on a piece of paper; they're cities or subway stations or fire exits. What makes a map a map, rather than just a drawing, is this correspondence.

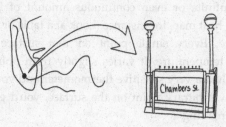

Chambers St.

Beyond that, there's a lot of flexibility as to what a map can be. The shape of the map doesn't have to reflect the actual physical shape of what it represents, as long as the correspondence is still there.

The dots or symbols don't even have to correspond to physical objects or places. They can indicate times, events, prices, just about anything, really. In the broader sense of the word "map," you just need a correspondence.

On most everyday maps, the meaning of each object is labeled directly on the picture. If a dot represents Buenos Aires, you write "Buenos Aires" next to it so everyone knows what's going on. On more complicated maps, that's not always so easy to do. If you're trying to assign meaning to hundreds or thousands of points, things get cluttered quickly. Labels aren't going to work.

There are other ways to draw maps that work better if you have an infinite or even continuous amount of information to show. A heat map, for instance. Look at a table or wall or any flat surface. Every single point on that surface has some particular temperature. It varies slightly from point to point, but if you had a very sensitive thermometer and you pressed it against any arbitrary point on the surface, you'd get an exact numerical reading.

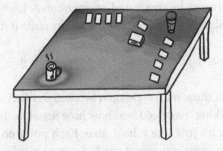

How could we draw a map to convey this temperature information? It's not going to be practical to label each point—we're dealing with a continuum of points here. So we have to get creative.

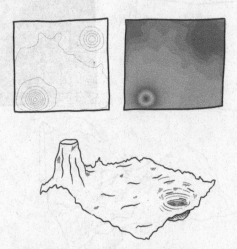

We can color-code the map so hotter points get lighter colors. We can draw contour lines that split the area up into regions of roughly equal temperature. Or we can add a temperature dimension: Hotter points are higher and colder points are lower.

Whichever version you prefer, these maps all present the same underlying information. What you're looking at is a correspondence between locations and temperatures. Each point on the table is assigned a value. Mathematicians write it like this:

map : {points on table} → {temperatures}

These same three map styles can be used in other situations too. A good hiking map needs to show how the altitude changes over a region. It's just like a heat map: Each point on the map corresponds to some numerical value. So we can color-code the altitude info, we can draw contour lines, or we can add a third dimension. (Here the three-dimensional map has a literal, physical interpretation.)

Topographic maps like this usually use the contour style, with a label on each contour line noting its elevation above sea level. But all three versions display the same underlying data.

map : {points in region} → {altitudes}

There's a reason we can use the same map styles for heat maps and topographic maps. In both cases we're mapping from a two-dimensional surface to a linear scale. Anything with this same basic structure will work the same way. You can show the linear scale visually, directly on the surface, in any of these three ways.

Any value that varies across a region can be mapped like this: annual rainfall, depth of a body of water, concentration of a pollutant, population density, etc. (The three-dimensional map of population density for a city would actually look a lot like the physical city skyline.) In all these situations, the data we're interested in is a correspondence between points in a two-dimensional space and points on a one-dimensional continuum. The general data structure looks like this:

$$\text{map} : \text{plane} \rightarrow \text{line}$$

But lots of things you might want to map don't fit this pattern, and these map styles won't work. Not everything fits on a linear gradient the way temperature and altitude do.

Like wind. Meteorologists need to map the wind, but "the wind" at a given location and time isn't just a quantity that can be color-coded. Wind has a speed, yes, but it also has a direction. A natural way to present this information is with arrows, where the length of the arrow indicates the strength of the wind.

This is a vector map. Each point in the space corresponds to a direction and strength. Vector maps are perfect for any situation involving a flowing substance, like the air that flows to create wind. The arrows show the direction and speed of flow at each point.

Next time you stir a cup of tea, pay attention to the flow of the liquid and see if you can imagine the vector map you're creating on the surface.

We're surrounded (literally) by flowing substances: The air around you is constantly shifting and churning. It's usually invisible to the human eye, but if you exhale in cold weather, or blow smoke or bubbles or dandelion seeds, you can briefly make out the outlines of the vector map your breath is creating.

In this case it's a three-dimensional flow, where each point in a three-dimensional space corresponds to a speed and direction.

You can also map flows of things that aren't material substances like air or tea. Heat flow is something engineers worry a lot about, and they analyze it using three-dimensional vector maps.

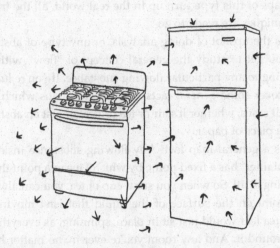

Vector maps can even be used to analyze flows of populations and resources across the globe.

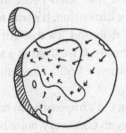

This would be a spherical flow, where each point on the sphere is assigned a vector value. You can have a map on any manifold.

Professionals in analysis tend to be specialized in one genre of map. "Real analysis" deals with linear quantities like temperature and altitude, and "complex analysis" is concerned with vector maps. Specialists in each camp know the ins and outs of their genre: how the maps behave, what they all have in common, what kinds of patterns and phenomena show up. Then, when maps of this type turn up in the real world, all the tricks and techniques are ready to go.

That's the upshot of doing analysis, or any type of abstract math. You get to study the general concept of "flow" without committing to any particular flowing substance. If you're lucky, you discover some general facts about vector maps which are true in all cases, whether it's air or tea or heat or just an abstract flow on a piece of paper.

Here's a general map fact: Any flowing substance inside a rigid container* has a fixed point, by which I mean a point that's not moving at all. So when you stir a cup of tea, you can always find a point on the surface of the liquid that isn't moving—where a tea leaf would just sit in place, spinning, as everything moves around it. And any room you're ever in, no matter how many fans are turned on, has some point of still air where a dust fleck would hover in place. (Assuming the windows are closed.)

This fact is called the "fixed point theorem" and it's been proven true in every dimension. It's true for a fluid churning in a two-dimensional dish and it's true for a gas churning in a three-dimensional bottle. If we lived in a world where we could build twelve-dimensional bottles and shake them up, it'd be true there too.

A related map fact: it's impossible to comb a hairy ball completely flat. If you try to take every point on a sphere and pick a direction for the hair to lie flat, you'll inevitably end up with at least one point of discontinuity, called a singularity or pole, where you get a cowlick or a part or a bald spot.

This fact applies not only to literal hairs, but to any time you try to assign a direction to each point on a sphere. Across the surface of the Earth, there's always at least one place where wind isn't blowing in any direction. In the oceans, we find singularities where the current doesn't flow in any direction, where trash collects and forms spinning islands. Even a tumultuous planet like Jupiter has to have at least one "eye of the storm" where the flow direction is undefined. This isn't just some observed pattern or coincidence of nature—it's a logical necessity, true even on planets we could never reach for billions of years. But it's only true for spheres: A hairy torus can be combed perfectly flat.

Maps, in this broader mathematical sense, are an incredibly versatile tool. They're used to analyze projections (like shadows and world maps), transformations (like rotations and reflections), quantities that change in time, geometric curves, states of physical systems, and plenty more. The functions you graph in high school are a form of map. The "stretching and squeezing" of topology is thought of as a way of mapping one shape onto another. Even the matchings from the previous two chapters are studied as discrete maps, where objects of one set "map to" objects of another. In pretty much any situation where one thing corresponds to another thing, mathematicians use a map.

Because when you look at things in the abstract like this, dusting off the specifics of a situation to focus on underlying dynamics, you start to realize there are only so many different patterns and structures out there. These patterns and structures are called mathematical objects, and thinking about them is called math.

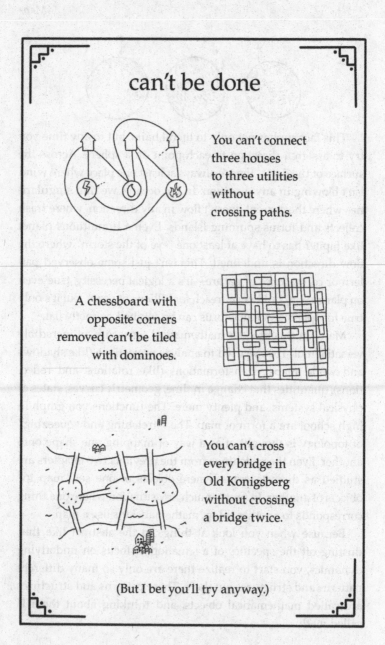

can't be done

You can't connect
three houses
to three utilities
without
crossing paths.

A chessboard with
opposite corners
removed can't be tiled
with dominoes.

You can't cross
every bridge in
Old Konigsberg
without crossing
a bridge twice.

(But I bet you'll try anyway.)

the pythagorean theorem

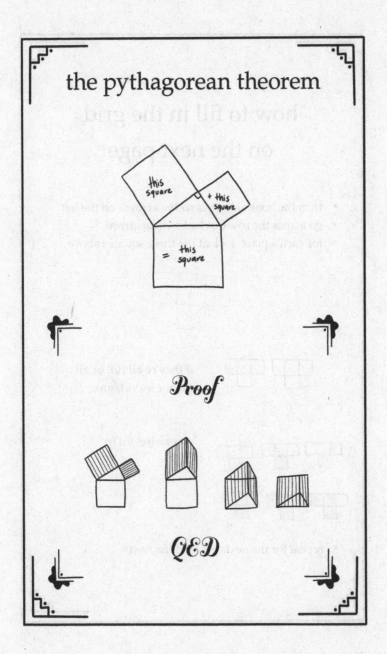

this square

+ this square

= this square

Proof

QED

how to fill in the grid
on the next page

- turn the book sideways so the arrow's on the left
- go across the row marked with an arrow
- for each square, look at the three squares above

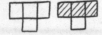 if they're all full or all empty, leave blank

 otherwise, fill in

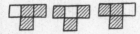

- repeat for the next row, and the next . . .

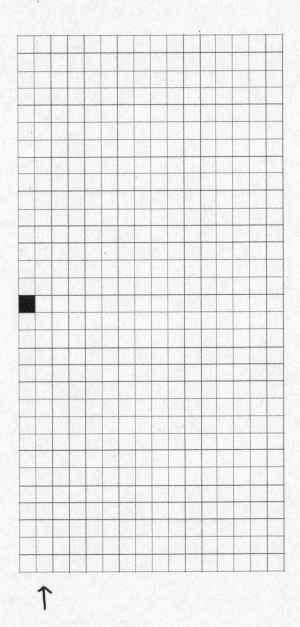

Algebra

abstraction

structures

inference

abstraction

*L*et's start completely from scratch. Math is about pure abstract objects sitting in empty space, and algebra is the purest, most abstract subject of all. We're not talking about the algebra you learn in school—hardcore mathematicians call that "school algebra" or "elementary algebra," and they mean it dismissively. What I want to tell you about in this section is called "abstract algebra." It's so abstract, it's not even about any particular type of object. It's about the very idea of objects and relationships between objects.

"Generalized algebra" is another term for it. When you generalize something, you make it less specific. Say you have a math problem with the number four in it. Four is a specific number. If you wanted to generalize the problem, you'd replace four with x, a placeholder for any number at all. Now you can't solve the problem the regular way and get a numerical answer, but you can sub in different values for x and see if there's a pattern to the answers you get. There usually is. That pattern is the solution to the generalized math problem. It's a solution that works *in general.*

Abstract algebra takes this idea to the next level: It tries to find a more general version of algebra itself. Instead of addition or multiplication, we use the symbol • as a placeholder for any operation at all. We sub in different operations—not just the classical four, but bizarre, never-before-used operations—and look for high-level patterns. And then it gets even worse: We abstract away the concept of number too, and now we're working with unknown operations on unknown objects.

It's hard to even talk about this type of algebra, because there's no particular thing to talk about. There are processes that algebraists carry out, systematic ways of moving symbols around on paper that turn statements into other statements. But each statement doesn't actually *mean* anything, or at least, it doesn't mean any one particular thing. Every symbol is a general placeholder for an infinite cast of possible replacements. So in a sense, each statement means a million different things at once.

It's a disorienting feeling. There's no solid floor to stand on, no clear reference point back to reality or even to what most people normally consider math. You can stare at pages of an algebra textbook for hours, flipping back and forth to try to remember what anything is even referring to. Whenever a proof or example does finally click, there's not usually a specific picture in your head so much as a sense of a pattern. "Something happened over here, and then a symmetrical thing happened over there, but it was flipped." There are clear relationships and structures, but no actual objects.

In order to think about this type of algebra, you need to get in the right mindset. You have to forget about real-world things like trees and chairs, and then you have to also forget about math-world things like shapes and numbers. You have to empty your mind, like you're preparing for a rigid and organized form of meditation.

So imagine this, if you can. Imagine you've never seen anything, heard anything, felt, smelt, or tasted, sensed or intuited, learned or known anything, ever. Imagine your eyes are closed forever and actually you have no eyes and you don't know what eyes are. You're just a disembodied consciousness floating in the void.

You have nothing to think about. Literally nothing: You don't know of any things. It's very boring. You have nothing to entertain you, and so you sit there completely blank for an eternity.

And then you receive a message, inserted directly into your mind. (Finally!) It says, "Something exists." It's a very basic message, but you're thrilled to have something to think about. *Something exists.* You don't know what it is that exists, but you know something does, and so you can give it a name, and you name it *g*.

Usually when you name something, the name has some kind of relevance to the thing. Not here. There's no etymology or onomatopoeia, and knowing that the thing is called *g* doesn't tell you anything at all about what the thing called *g* is. It's just a name, a symbol, that you use for ease of reference. You can say things like, "*g* exists." You can even draw a conceptual diagram of everything you know exists in the world.

• *9*

But don't read too far into it: The thing is not really a *g*, and it's not really a point. You're just sketching the abstract idea of the thing that you call *g*.

Now you're bored again. You've done pretty much every-thing you can do with this one object that you know exists, and it turns out one generic thing existing isn't much more interest-ing than nothing existing at all. So you go back to utter blank-ness, wishing you had thumbs to twiddle, for another few eons.

Then, thankfully, the messenger comes back with a new message. "Another thing exists." Great news! You give this new thing the name h and you update your little schematic diagram.

$$\bullet \, g \qquad \bullet \, h$$

But once again, that's about all you can do.

No matter how many new things you hear about, all you'll be able to do is add new names to your list of names, add new dots to your dot diagram, and then go back to nothingness. If someone asked you, "Does h exist?" you could tell them that yes, it does, but you still don't know anything beyond exactly what you've been told by the messenger. You can't figure out new facts on your own. You can't ask questions and wonder about the answers. The world consists of a list of unrelated objects, and there's not much you can do with that. Boring!

In order for anything even remotely interesting to happen, you need to know not just whether things exist but how things relate to each other.

So consider this. The messenger comes back and gives you a message with a new twist to it. "There are five things that exist, and each of these five things has a partner-thing that also exists." Alright, let's think about that. What could this message be describing?

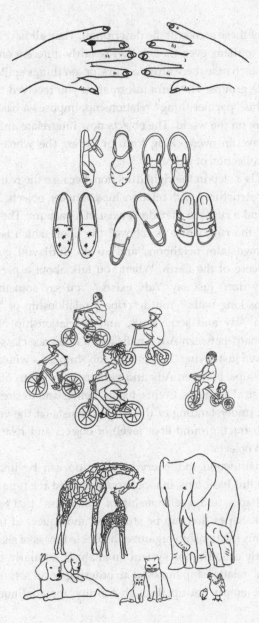

All of these scenes fit the description. They all have a similar pattern to them, even though they're fairly different on the surface. In each case there are five pairs, or ten things split into two alternate groups. The extra information you received this time, about this "partner-thing" relationship, imposes a basic sort of structure on the world. The objects now interrelate and coexist. They have an overarching form or order. The whole is more than a collection of parts.

This is a step in the right direction, because the real world is densely structured with relationships between objects. There's a couch and a rug, but they don't exist in a vacuum: The couch is "above" the rug, which is "above" the floor, which is "above" your downstairs neighbors, and so on until you get to the molten core of the Earth. When you talk about a person, you probably don't just say "Ady exists." You say something like "Ady has long nails." You describe a relationship of "having" between Ady and some nails, and a relationship of "being longer than" between Ady's nails and a reference class of other nails. Even just saying "Ady is a person" implies a whole suite of relationships, between Ady and various body parts, other people, physical locations, events, habits, beliefs and desires, and so on. Our understanding of the world consists (at the very most basic, abstract, ground-floor level) of objects and relationships between objects.

In math-world too, everything we do can be understood through this basic lens. In topology, we looked at a type of object called shapes, and a relationship of "sameness" that applies to any two shapes that can be stretched and squeezed into each other. This relationship organizes a jumbled mess of shapes into an orderly classification system. In analysis, similarly, the "bigger than" relationship imposed an ordering on all sets of things, from the empty set up through infinity, the continuum, and beyond.

But I said we were leaving the real world and math-world behind. So let's forget about all that for now and go back to your consciousness floating in the void, and those five things that exist with their five partner-things that also exist.

You want to name the objects, and you want to make a dot diagram. It doesn't seem quite right to give ten random different names to the objects, because that doesn't reflect what they are. You *could*, since the names don't really mean anything, but to make life easier for yourself you might as well choose names that reflect the order of the world you're describing. Same with the dot diagram. You could scatter the points randomly, but you're better off if you indicate the partner-things.

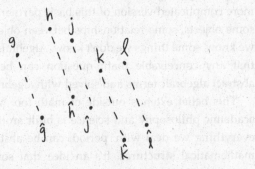

This is one of the simplest possible worlds that has any organized structure to it. That's good, because mathematicians like things simple. That's the point of abstraction: It lets us investigate the nature of order and form, without getting tied up in the arbitrary specifics of any particular situation.

So what exactly do we learn about order and form? What makes this structured world, with the partner-things, qualitatively different from a collection of unrelated objects?

For one thing, this world can be talked about in a way that the previous worlds couldn't. We can say, for instance, "*g* is the partner-thing of *ĝ*" and "*h* and *j* aren't partner-things to each other" and "$\hat{k}$ has a partner-thing but it isn't g or the partner-thing of h." Before, without relationships between objects, all we could say was that things exist. As in the real world, relationships are the substance of language.

This also means that we can ask questions and search for answers. Like so: "What's the partner-thing of g?" Or, "Is there any object that is not the partner-thing of its own partner-thing?" These questions are easy to answer, because the world we're dealing with still has hardly anything going on. But for the first time, we're in a situation where there can be things to discover.

Backing out for a second, the big idea behind abstract algebra is that everything we deal with in math is essentially a slightly more complicated version of this basic partner-world. You have some objects, some relationships between objects, some things we know, some things we don't know. Algebraists are convinced that any conceivable math question can be translated into abstract algebraic terms and solved with algebraic tools.

This belief extends outside of math too. Much of Western academic philosophy and science is built around the idea that everything we deal with, period, can be abstracted to simple mathematical structures. It's an idea that sounds crazy, and quite possibly *is* actually crazy and wrong. At the very least, it's a powerful idea that has helped people understand the workings of nature and build new technologies.

Because the partner-thing example is still too basic to be interesting, I want to give one more toy example of an abstract structure. Forget everything you know one last time. Ready? Here comes your message. "There are three special things, let's call them 'wugs,' and each possible combination of 'wugs' is also a thing that exists."

What could the messenger be describing? A simple naming system we can use would look like this:

g h j gh hj gj ghj

But how should we arrange the points? What's a visual schematic that captures this structure? There are several that would work, but here's a good one: the vertices of a cube.

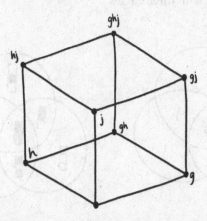

Take a minute to look at this diagram and appreciate why it works as a representation of the structure. The near corner represents the empty object—the combination of no wugs. And then, each wug corresponds to one of the three dimensions. To add the wug, travel in that direction.

Structure diagrams like this always have nice symmetries and patterns to them. See how opposite corners of the cube have opposite names? That's a sign we're capturing the underlying structure well.

Another world with this exact same structure to it would be the collection of all three-bit binary strings, which is to say, the possible states of three on-off switches.

$$g \quad h \quad j \quad gh \quad hj \quad gj \quad ghj$$

Here, the wugs are the switches, and the empty object is all three off.

Another presentation of the same underlying structure: a Venn diagram with three circles.

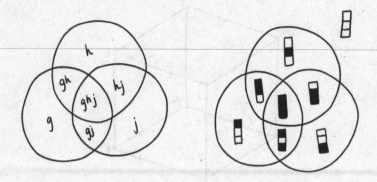

One last system that also fits into this same pattern: colors.

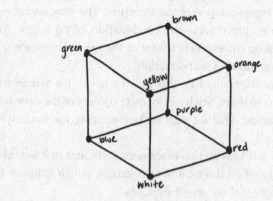

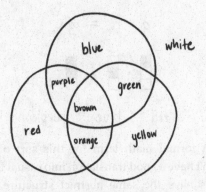

Curiously, the factors of thirty also fit this exact same pattern. I'd show you what that looks like (it's pretty neat) but I promised I wouldn't use any numbers, so I guess you'll have to work that one out for yourself.

Specifics aside, the general point I'm trying to make here is: The same underlying abstract structure can manifest as any number of superficially different systems. The specific objects in each case are wildly different, but the relationships between them are exactly the same.

	wugs	combine	nothing	everything
ghj - world	letters	append		ghj far corner
cube	dimensions	→↑	near corner	▮
bit strings	positions	overlay	目	
venn diagram	circles	overlap	outside	center region
colors	primary colors	mix	white	brown
factors of thirty	primes	multiply	one	thirty

Another way to think about this "same abstract structure" equivalence: Any sentence you can say about one set of objects is still true when translated word by word into another.

$$g \cdot h = gh$$

red • blue = purple

There's a formal math term for this sort of equivalence, which doesn't have a good translation into casual English. When two systems have the same abstract structure, we say that they're isomorphic to each other, *iso* meaning "equal" and *morph* meaning "shape" or "form." The colors and the bit-strings and the cube corners are all isomorphic—they have the same conceptual shape.

When you read the word "isomorphic" in a textbook, the author is probably being very precise with the definition, describing two systems that have the exact same structure, with no differences at all. But also, mathematicians sometimes call things isomorphic in real life, and then we usually mean it in a rougher, impressionistic sort of way. You might say that Uno is isomorphic to Crazy Eights or that *The Lion King* is isomorphic to *Hamlet*, even though those aren't true in the most literal, mathematical sense.

To an algebraist, isomorphism—the state of being isomorphic—is the pinnacle of elegance and beauty. Two unrelated situations that secretly have all the same underlying dynamics? *Gorgeous.* The world has been simplified down one notch. What used to be two different problems or, as the case may be, a hundred or infinity different problems, has been reduced to a single problem. Our understanding is deepened. (At least, that's my impression of an algebraist.)

We saw this kind of abstraction/reduction process in action a few chapters ago; I just didn't call it that at the time. Think back, or flip back, to Hotel Infinity. Adding a new guest to a full hotel is a situation with the abstract structure of "infinity plus one." Adding one new object to a bottomless bag of objects is the same thing: infinity plus one. Once you understand the dynamics of "infinity plus one" in one scenario, the exact same logic applies to any isomorphic scenario.

After all, think about it: What does "infinity" or "one" even mean? One *what*? These ideas are abstractions. One duck, one hair, one drop, one minute—the concept applies across a million different cases, but it doesn't mean anything on its own. It's a placeholder term for a recurring phenomenon. It's an abstract object, a pure mathematical object.

What exactly is a "pure mathematical object"? Do these things really exist or are they just figments of our imagination? These are questions that philosophers of math argue about. Some people believe that mathematical objects do exist, in the most literal sense, in some far-off universe of pure abstraction. They believe we catch glimpses of this simpler world when we study math. They believe this pure math universe, the "Platonic realm," is more fundamental and more beautiful than our world, less arbitrary, less influenced by happenstance.

I don't know about all that, but it's a useful way to think about abstract objects. Whatever the *ghj*-structure is, we can imagine it sitting in that blank, empty math-world. We can't know what it looks like, just like we can't know what a pure "one" looks like. We can see all these various shadows it casts onto our world—the cube, the Venn diagram, etc. But the *object* we're talking about? This skeleton of an idea, this abstract algebraic shape that I'm trying to communicate from my head to yours? This is just a *thing*. A structure. We can give it a name, and so we name it *Z*-two-cubed.

Yes, of course: Mathematicians have made it a mission to find, name, and classify every possible abstract structure.

structures

$\mathcal{D}$on't worry: We're not about to go through an exhaustive classification of all possible abstract structures. Who has the time? There are a lot of them, which makes sense because "structure" is a very broad concept. It's not like when we classified manifolds, going in order by dimension and jotting down a few shapes each time. Classifying algebraic structures is a project closer in form to classifying all species of life on Earth. It happens in layers: There's the top-level domain of "structure," but then there are a dozen or so recognized categories of structure—fields, rings, groups, loops, graphs, lattices, orderings, semigroups, groupoids, monoids, magmas, modules, and then a whole bunch that we just lazily call algebras. Each of these categories has subcategories, which have sub-subcategories, which can be further sliced up by their properties and characteristics. It's a pretty imposing taxonomy.

So, with no aim of catching them all, I want to take this chapter to show you a sampling of some of the structures out there.

I'm hand-picking the types of structures that appear more often in the wild, but please remember: Professional mathematicians are not especially concerned with what's common or useful outside of math. Algebraists study structures they find interesting or elegant, whether or not they have any known manifestations in our world.

Sets

A set is the simplest structure of all. It's so simple that some people don't even consider it a structure. It's a collection of objects, with no other relationships or properties.

Here's one example of a set. This set is called *two*. We can't actually look at the set (it's an abstract object with no specific form), but here are various real-world scenarios with this structure.

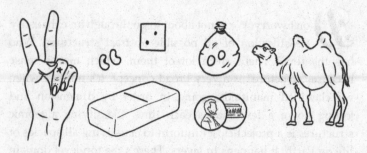

There's no "correct" way to look at or draw a set—there's no single "correct" way to look at any kind of structure.

This set, two, is one of many different sets that exist. Finite sets are very easy to classify. Every finite set is isomorphic to one of the following:

Infinite sets are a little trickier, to put it lightly.

Graphs

A graph is similar to a set, but with additional structure. In a graph, some objects have a special relationship with each other. We can draw the objects as dots, and the relationship as lines between the dots.

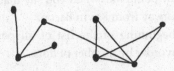

You can look at this as the structure of a social network, where each dot is a person and each line is a friendship. At least, this is how social networks are structured on websites like Facebook or LinkedIn, where friendship is a binary and it always goes both ways.

You could also pick something more precise than "friendship" to draw the links on this graph. You could connect two people if they've ever spoken to each other while making eye contact. You could only connect people who have kissed. You could connect two people if and only if they've acted together in a movie listed on IMDb.

Common questions asked about graphs include: How densely interconnected is it? How segregated into different cliques? Does it cut clean into two subgraphs, with no connections between them?

Can it be drawn without any lines crossing? Are there any lone dots without any connections? Which object has the most connections? Which object has the most "friend of a friend"

two-step connections? Which objects are the most central, meaning, which are the fewest steps away from all other objects?

If it's true that you're at most six degrees of separation away from all people on Earth, that means the "diameter" of the human social graph is six. You can also calculate the "radius" from a particular point, like when you say that every actor is at most four steps away from Kevin Bacon.

Here's the beginning of the list of all possible connected graphs, going in order by number of dots:

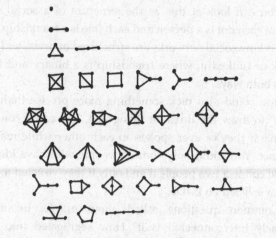

Weighted graphs

In real life you might imagine that friendship isn't a binary so much as a continuum. For any pair of people, the relationship can vary between zero (never met) and infinity (inseparable). That has the structure of a weighted graph.

You can't even begin to list all possible weighted graphs. There's a continuum of different options, even just for a two-person weighted graph.

Directed graphs

A directed graph is similar to a graph, but instead of symmetric lines there are one-way arrows.

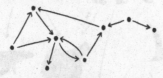

People on Instagram or Twitter form a directed graph, because you can follow people who don't follow you back.

The structure of the internet itself is a directed graph. Each page is a node of the graph, and each arrow is a link from one page to another. When you click from page to page, you're following a chain of arrows. Most modern search engines use graph theory to sort the search results they show you, pushing pages to the top if they have more links pointing into them. (They also use other considerations, like advertising, and how closely it matches your search query.)

The game rock-paper-scissors can also be modeled as a directed graph with three nodes.

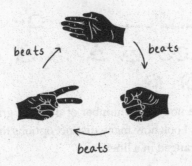

A common question about directed graphs is whether they have cycles: If you follow a chain of arrows, will you ever end up back where you started? Rock-paper-scissors has a cycle, while a typical food web doesn't.

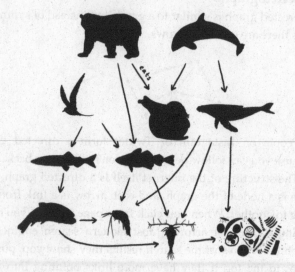

Here are the first bunch of connected directed graphs.

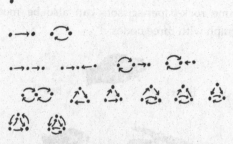

After three nodes, the number of directed graphs blows up really quickly. Look how many distinct options there are just for four nodes arranged in a line:

Game trees

There's a common type of two-player game that mathematicians love to think about. Checkers, chess, tic-tac-toe, Go, Connect Four, and Reversi, aka Othello, all fall into this category. These are games with no elements of luck, where two players take turns making moves, with complete information, and at the end one person wins or it's a draw. These are called combinatorial games, and they can be studied as a type of structure.

Let's look at tic-tac-toe. Each possible board position is a dot or node, and there are two types of arrows: moves that X can make, and moves that O can make.

Using this "game tree," you can follow any one game of tic-tac-toe as a path down the tree that alternates X arrow, O arrow, X arrow, O arrow.

Every combinatorial game—checkers, chess, etc.—can be turned into a game tree like this. The game tree for a game like Go, where there are hundreds of legal moves on every turn, would be insane to actually write out on paper. But computers that play board games are programmed to search through game trees in order to find good strategies.

You know how in tic-tac-toe it's always possible to force a draw, if both players play well? An interesting fact from combinatorial game theory says that *every* combinatorial game is like this: It's either a forced win for one player, or it's a forced draw. If both players play optimally, the game is decided from the very beginning. In practice, we still don't know what the optimal strategy is for games as complicated as chess or Go. But in theory, every complete-information game without luck* is "solvable."

Take any combinatorial game and write down its full game tree. Let's call the two players X and O. Any board position where the game is over, color that node green if X won, red if X lost, and gray if it was a tie.

Now we can color in the rest of the tree, not just the end positions. If there's a position where it's X's turn and there's an X arrow pointing to a green (win) position, color it green, because *X* can play a winning move. If all the *X* arrows point to red (lose) positions, color it red too. If there are *X* arrows pointing to both red (lose) and gray (tie) positions, color the node gray, because *X* can choose to tie.

Keep coloring positions this way, going up the tree until every node is colored.

What color is the starting position? If it's green, *X* can force a win. If it's red, *O* can force a win. If it's gray, the game is a draw with best play.

Checkers and Connect Four have both been solved, using powerful computers that search exhaustively through every branch of the game tree. (Checkers is a draw if both players play perfectly, and Connect Four is a win for the first player.) Still, it's unlikely that any human could ever memorize the perfect strategy for all possible situations, so the game is still interesting to play in practice. Chess and Go haven't yet been solved. Some top chess players have suggested that chess is ultimately a forced draw with perfect play.

Family trees

A family tree is also a graph-like structure, consisting of nodes and connections. But now each connection is like a split-tailed arrow, indicating the "parents of" relationship.

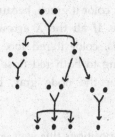

You can specify that each parenting arrow needs to have exactly two parents, or you can allow for other family structures. Here are the first few family trees, allowing for any number of parents.

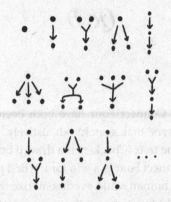

I'll say this again, since it's important: These dot-and-arrow sketches are just one convenient way of representing structures. Structures themselves have no earthly form. Algebraists often just describe structures in math language, without pictures, like this:

A *family tree* is a set S equipped with a parenting relation $\{(P_i, x_i)\}$ which holds between parent-sets $P \subseteq S$ and children $x \in S$.

Symmetry groups

I have to talk about symmetry groups because algebraists are obsessed with symmetry. Theoretical physicists are obsessed with symmetry too, and these days many theoretical physicists have to study group theory on the side, to get practice working with different symmetries.

You've probably noticed that different shapes, patterns, and objects have different forms of symmetry. There's flip symmetry, rotational symmetry, translational symmetry, and dilational symmetry.

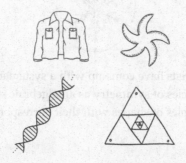

Each of these types of symmetry has so, so many different subtypes. For instance, rotational symmetry can be discrete or continuous:

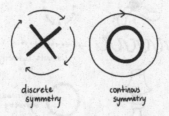

An object can also have rotational symmetry along multiple axes of rotation.

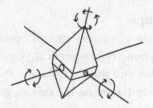

You can also have rotational symmetry and other types of symmetry at the same time.

Group theorists have come up with a systematic way to represent each species of symmetry as an algebraic structure. Here are some examples of shapes with their corresponding symmetry groups.

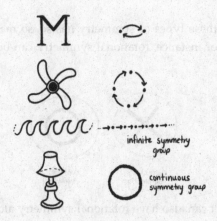

infinite symmetry group

continuous symmetry group

These examples each have one type of symmetry. The group structure is a little more complicated for objects with multiple types of symmetry. For instance, here's the symmetry group of a square. It has two types of arrows, one indicating a left-right flip and the other indicating a quarter rotation clockwise.

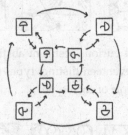

Wallpaper groups

This last one is a subcategory of symmetry groups. An object or pattern has wallpaper symmetry if it can be used to tessellate the entire plane. These designs all have wallpaper symmetry but no other form of symmetry:

Whereas these designs have wallpaper symmetry and four-sided rotational symmetry:

And these designs have wallpaper symmetry, flip symmetry, and six-sided rotational symmetry:

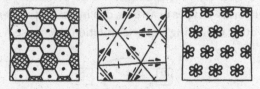

A beautiful and curious result of abstract algebra says that there are exactly seventeen distinct types of wallpaper symmetry. Here's one sample of each type.

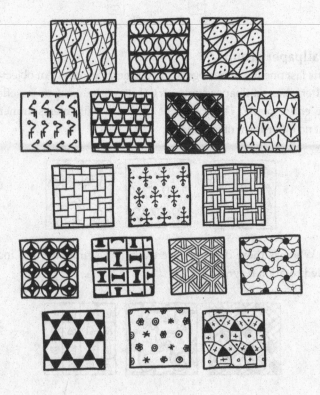

I'll stop listing structures now, but keep in mind there are many, *many* more categories not on this list. Algebraic structures can be used to model just about anything with patterns and regularities: English syntax, codes and cryptography, music theory, Rubik's cubes, anagrams, particles, supply chains, polynomials, juggling, you name it. Everything on a computer or phone is stored in memory as a "data structure"—another type of algebraic object. There's even a very meta branch of algebra called category theory, which studies *categories of structure*, looking for patterns and relationships between all these categories.

After all, an algebraic structure is just a set of interrelated things. It's a pretty versatile tool. That's why many algebraists are convinced that, if you really want to, you can represent anything in the universe as some type of abstract structure.

inference

*B*ack to the real world for a minute. Think about a city. Think about a million people going about their days, interacting with each other. What types of relationships exist? What is the structure of this network of people? It's not simple—none of the structures I showed you has anything near this level of nuance. Think about all the true statements you could make about people in a city: "Che has cousins in France." "Deb and Max went on a weekend trip together last October." We're a long way from "red • blue = purple."

We do live within structures. They're too complex to analyze with algebraic precision, but they're structures. What you eat, where you sleep, who you love: These facts exist within local, regional, and global networks of trust, trade, power, labor, coercion, tradition, accountability, and so forth. We don't need to know exactly how this all fits on a page, what all the arrows and dots look like, to get that on some level this is all one big system of interrelated parts.

Being inside a structure is very different than looking down at one drawn on a piece of paper. From the outside, everything is known. You can see all things and all relationships between things at once. Down here inside the system, we can only see little bits and pieces. We know things about the people we interact with, and we get hints as to what's going on beyond our local networks. That's about it.

From this starting point of limited information, we manage to figure out a lot about the world. We pick up on patterns and fill in the blanks. We use common sense and logic to parlay the little snippets we're given into new, useful knowledge. How do we do that?

We're constantly making inferences, but it's worth stepping back to appreciate what a remarkable feat that is. You take something you know—something you're told, or something you can see for yourself—and magically shake it around in your head and turn it into some new thing that you now also know. You see a single street sign and suddenly you know which direction you're facing and how to get to the park. You're told the sea levels are rising, and now you also know that people on islands are in danger. The more complex the system, the more impressive it is that we can deduce so fluently.

What's going on mechanically when you jump from one truth to another like that? When can you safely make inferences, and when will you accidentally jump to conclusions that aren't true?

Mathematicians study inference out of both interest and practicality: Maybe if we can get this process down to a science, we'll be able to formalize and automate it. We'll be able to learn everything there is to know about a system just by plugging in some basic facts and pressing "infer." (At least, that's the dream.)

Unfortunately, the real world is complicated and difficult to systematize. There's a lot going on and there aren't clear rules. So what do we do? We abstract! We consider a way, way simpler world and investigate how inference works in that world. By testing the waters with a bunch of oversimplified scenarios, we can get a sense for how inference works in the general case.

So let's get to it. What's a simple system we can make inferences about? Good thing we spent the last chapter running through example after example of basic structures—we can use one of those. Let's do a family tree. How does inference work when the system you're reasoning about is a family tree?

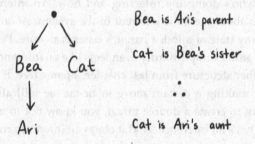

Say I tell you, "Bea is Ari's parent," and you've already heard that "Cat is Bea's sister." Based on this, you can make an inference: Cat must be Ari's aunt.

This particular inference is about Ari, Bea, and Cat, but it's clear that the same pattern of inference will apply in other cases. If Bea has another child named Zeb, then you know Cat is Zeb's aunt too. If Cat herself has a parent who has a sister, then Cat has an aunt. Across all conceivable family trees, there's a general inference rule about aunts that always holds.

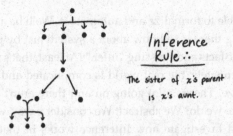

Inference
Rule ∴

The sister of *x*'s parent
is *x*'s aunt.

Sure, it might seem like overkill to call this a "rule." It's not like you have to consult some official rulebook whenever you want to determine aunthood. Most likely, you just knew intuitively that Cat is Ari's aunt.

Our goal here isn't to understand literally how human brains reason about systems. That's a problem for psychologists and neuroscientists. Our interest is in the inferences themselves: We want to know which types of inferences are legitimate, regardless of who's doing the inferring and how. An inference rule tells you about the logic inherent in the system. At any time of day, in any state of mind, a parent's sister is an aunt. Period.

This aunt example hardly even feels like an inference, so let's try another structure from last chapter: a game tree. If you realize that making a certain move in tic-tac-toe will allow your opponent to create a double threat, you know not to make that move. That's an inference. And it obeys an inference rule.

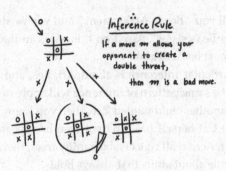

Inference Rule

If a move *m* allows your
opponent to create a
double threat,

then *m* is a bad move.

One more super simple example: an ordered set. If you know the sun is older than the Earth and the Earth is older than the moon, then you also know, naturally, that the sun is older than the moon.

$$\text{Sun} > \text{Earth} > \text{Moon}$$

Inference
∴ Rule

If $a > b$
and $b > c$,
then $a > c$

(I didn't include ordered sets in our catalog of structures last chapter, but it's a pretty intuitive concept, no?)

In all these elementary systems, we reason according to inference rules. A system permits certain patterns of deduction, and we can write down those patterns as inference rules.

Each system has its own particular set of inference rules, reflecting the particular structure of knowledge in that system. When you reason about a game of checkers, you're necessarily going to be following a different set of inference rules than when you reason about navigation or social movements. The examples we deal with in math are always bare and simplified, but you could imagine that even a more intricate, real-world system might have a consistent logic to it that can be written down as inference rules.

Across all systems, the basic form of an inference rule looks something like this:

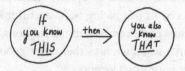

Inference rules are a simple but tremendously powerful thing. Once you've written down a list of inference rules for a system, you've found a key for unlocking new caches of knowledge. It's a chain reaction: You use *A* to deduce *B*, and then you use *B* to deduce *C*, and then *D*, and *E*. . . . You may then find that *A* and *D* being simultaneously true necessitates that some other statement *P* is also true, which sets off a new chain of inferences, which further combine and multiply with other things you know. Now when someone tells you one new fact, *boom*, it branches out into a thick web of interrelated truths.

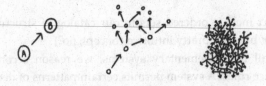

A lot of algebra—abstract algebra as well as the school kind—amounts to a careful application of strict inference rules. Think about a school algebra problem where you have to solve for *x*. They give you an algebraic expression to start with, which represents some fact about a system. Then you start applying inference rules: "If this is true, then it'll still be true when I add one to both sides." Each step is a basic inference and somehow, by the end of the process, *voilà*, you've learned what *x* is.

Sometimes *x* is just *x*. In that algebra homework, there's not going to be any higher meaning to the end result, so it can feel like a whole load of nothing. But these same formal inference procedures can also be used in real-life scenarios, and they genuinely do succeed in producing new, useful information. Your GPS, to name just one example among millions, measures your distance to three satellites and then uses geometric inference rules to deduce your exact location.

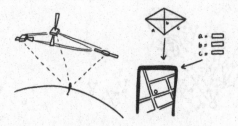

Our world today is saturated with these rigorous, systematic processes of deduction. They're in our machines, they predict the weather and issue safety warnings, they manage transportation networks and trade networks and government programs. Businesses use algebra to maximize profits and advertisers use algorithms to predict (with devious accuracy) what we'll want to buy. Theoretical physicists actually used abstract algebra to predict the existence of subatomic particles called quarks, which was later confirmed in experiments. And it's not just a new-age thing: Similar systems of formal inference have been used by most world cultures throughout history to predict the motions of stars in the sky.

I might go as far as to say that mathematicians love the idea of formal inference rules a little *too* much. I mean, it's not hard to see why. A small core of truths can explode out into a dense network of knowledge? Amazing! Think about all you can discover just by sitting down with paper and pen! You can learn new truths of the universe by obeying rules and moving symbols around. It's like you cross-multiplied and it taught you about the nature of reality.

But somewhere along the way, people got intoxicated on this idea and really let things get out of hand. They started running it in reverse. They thought: If inference rules turn less knowledge into more knowledge, then maybe you can take any bundle of facts and reduce it back down to a small core of fundamental truths that imply everything else.

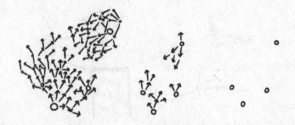

In simple mathematical systems, this reduction trick does seem to be possible. For instance, mathematicians were able to reduce all facts of arithmetic to the following five statements:

Zero is a number.

•

If x is a number.
the successor of x is a number.

•

Zero is not the successor of a number.

•

Two numbers with the same successor
are the same number.

•

If a set S contains zero,
and S contains the successor of every number in S,
then S contains every number.

This is called an axiom system. Everything you can know about every whole number, about multiplication and primes and all that, can (in theory) be deduced from these five axioms.* Which, you have to admit, is pretty impressive. It's a concise and elegant starter pack for arithmetic. You can walk around and say, "I know these five simple things, and therefore I know everything there is to know about arithmetic." That's a powerful feeling, like creating a universe by rubbing a few sticks together.

In practice, you'd never actually use these five axioms to prove something new. Even basic facts of arithmetic are incredibly hard to prove if you have to start all the way from axioms, without using any other knowledge at all. I mean, look how hard it is to prove that one plus one is two:

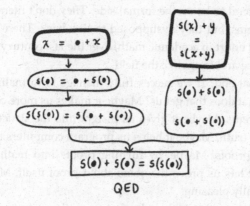

This kind of proof is called a formal proof. You start from axioms, and you're only allowed to use inference rules. You can't rely on intuition or common sense, only inference rules. Yes, you can use facts that have previously been proven from the axioms, but everything has to ultimately link back to the axioms. The goal of this kind of proof isn't to be convincing in the rhetorical sense—formal proofs are hardly even readable most of the time!—but to locate your claim within this rigid and precise system of accepted truths.

This is a controversial topic in the math community. To what extent should we use formal proofs? People generally think they're more trustworthy than the informal, intuitive sorts of arguments we've been using in this book. They follow strict rules, so they're not as vulnerable to human error. But many people, especially students, find them confusing and off-putting. They read like a foreign language. They're often presented as concisely as possible, without an explanation of why each step is taken or what the overall argument is.

Whichever side you take, one thing is clear: Formal proofs have taken over. There's still room for intuition in classrooms and informal settings, but the proofs in textbooks and math journals tend to be on the formal side. They don't literally start from axioms, but they're supposed to link back. There's been a concerted effort in academic math, over the last century or so, to axiomatize and formalize the field.

To what end? If we successfully manage to formalize every proof, what does that get us? Maybe it makes us more certain of our theorems. Maybe it gives us insight into the structure and nature of truth. Maybe it helps us program computers to generate new proofs. Maybe, by turning proofs into mathematical objects, it lets us prove theorems about proof itself. Maybe it's aesthetically pleasing.

There's one thing, though, that formalism won't get us. It won't let us divide the world into "provably true" and "provably false." That was a big driver of the original push toward formalism: people thought it would give us a systematic, objective way to determine whether any statement is true or false. Then that hope came crashing down, dramatically and permanently.

Remember when I said there's a lesser-known third status between true and false? Now I'm ready to tell you about it.

two math games

"Coin game"

→ put some coins on a table
→ two players take turns
→ on your turn, take one or
   two coins
→ whoever takes the last
   coin wins

Difficulty to find the winning strategy:
EASY-MEDIUM

"Puppies and kittens"

→ start with two piles of
   different sizes
→ two players take turns
→ on your turn, either:
a. take any number of
    coins from a single pile
b. take the same number of coins from both piles
→ whoever takes the last coin wins

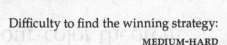

Difficulty to find the winning strategy:
MEDIUM-HARD

the four-color theorem

Every map can be colored with four colors so that no neighboring countries are the same color.

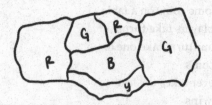

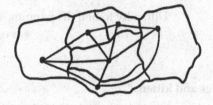

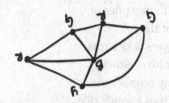

Every graph without crossings can be colored with four colors so that no linked nodes are the same color.

the four-color theorem

how to draw a dodecahedron

- draw a pentagon with all sides and angles equal
- weave another identical pentagon upside-down (lighter)
- draw a short line from each corner, away from the center
- connect with ten straight lines

how to draw an icosahedron

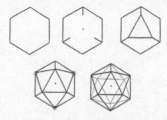

- draw a hexagon with all sides and angles equal
- draw a short line from three corners, toward the center
- connect with a triangle
- connect the other three corners to the two near triangle corners
- optional: repeat all steps upside-down (lighter)

Foundations

a dialog

So: some things can be proven true, some
things can be proven false, and—

>> Wait a minute, wait. Something's fishy here.

Mm?

>> Well, earlier on we were going through proofs of
>> things. We'd make a claim, and then there'd be a
>> convincing argument for why that claim is true.
>> But now you're just saying things like "It's been
>> proven" and "As it turns out." What's with that?

There's a lot to get through! We can't go over every
single proof. I hand-picked a couple nice ones to
include, but a lot of proofs are tedious, and I didn't
want to bore you with long, case-by-case details.

>> But you've seen proofs of all these things?
>> And they were totally convincing?

Most of them, yes. Some of the proofs are very
beautiful and I can show them to you if you like.
They're very convincing. Bulletproof. Some of them
I admit I haven't personally seen, but I know
they've been proven. They're cited all the time and
generally accepted to be true, to be fully valid
proofs.

>> Valid by whose judgment? I'm not trying to ques-
>> tion your judgment or anything, I just mean people

disagree about things all the time and what's convincing to one person isn't always convincing proof to another person. You don't always get unanimous juries. So it seems like you'd have people—even all really smart, mathy people—they'd disagree sometimes about what counts a valid proof or not.

Sure, people disagree about things, but we're not talking about a courtroom here. No one's getting paid to argue a certain side. We're all working together to figure out what's true.

Still . . .

And besides, math is way less complicated than the stuff people normally disagree about. I mean, we're talking about basic shapes and structures here. There's no "he said, she said." There aren't that many moving parts.

Sure, but even some of the proofs you did show me, I was a little puzzled by the argument. I think I get it, but it's not the most obvious thing in the world. And these long, complicated proofs that you're not even showing me, some of them you haven't even seen . . . How am I supposed to trust that? You see how that's suspicious, right?

Right, totally.

I mean, has there ever been a "proof" that everyone accepted and then it turned out to be wrong.

Well . . . yes, actually. But that's really the exception
and not the rule. We have a great track record with
peer review and everything. We're very strict about
what counts as valid proof.

But it has happened?

Yeah, but really only once or twice
in any kind of significant way.

What theorem was it?

It was the Four-Color Theorem. Says that if you
have a bunch of countries on a fictional world map,
and you want to color them so no two adjacent
countries are the same color, you can always do it
with at most four colors. Any map at all.

And that's not actually true? It was wrong?

No, no, it's true! But there was a proof someone
found, a long time ago, a beautiful and relatively
simple proof, and it passed peer review and everyone was satisfied.

And then someone found a flaw.

Right, the proof was invalid. There was one case it
missed. And people tried to patch up that one case,
but no one ever succeeded, and so it went back to
being unproven. The Four-Color Conjecture.

Well, wait, then how do you know it's true?

Because now it *has* been proven! Using computers.
A totally different proof, with some heavy-duty
graph theory, clocking in at a few hundred pages.

> But see, you're still acting like this new computer
> proof is the ultimate decider of what's true. What if
> it turns out to be wrong again? You need to link
> back to something real, otherwise you're just chas-
> ing your own tail. "Math says x is true, and it's
> right, x is true, because math says so."

So you think we're all just wrong? Everyone, all
these mathematicians, we're all systematically
wrong in the same way? What are the chances
of that?

> It's happened before, hasn't it? That actually hap-
> pens a lot historically, where everyone's systemati-
> cally wrong about the same thing, it's just a thing
> everyone's told is true and no one really thinks to
> question it and you're ostracized or shamed for
> thinking it might be wrong.

Okay . . .

> And I'm not saying all this stuff is *wrong*, like flat-
> out, objectively wrong. It's just all contextual, no?
> Culture clearly influences what we think is true or
> obvious. So the math community has some consen-
> sus about what proofs you'll accept as valid. Great!
> You can follow those rules, I won't stop you. I just
> don't see why everyone else has to also accept it all
> at face value.

Okay, granted, context is important, and people can be systematically wrong as a group. That definitely happens a lot for things like politics and morals, and sure, it even happens for science. There are tons of things that used to be scientific consensus, things like leeches and yellow bile, and also entire fields of science that were basically just people writing down racist political ideology in the language of science.

Exactly!

But I think math is different. Really!
Let me at least say why.

Try me.

The thing about math is that it's never been just one isolated culture doing math and self-reinforcing and punishing dissidents or whatever. As far as we can tell, every single human culture has independently come up with math. "It's the same in every country," as they say.

Okay, good point.

Things like astronomy, geography and navigation, counting and record-keeping, geometry, architecture, some form of money and gambling, some form of logical reasoning, irrigation, measurement and construction . . . All these tools were developed separately by pretty much every society we know of.

Right, you wouldn't expect us to all get counting wrong. That'd be hard to coordinate.

And sure, maybe it's knots in a string over here and tally marks over there, but it's all the same ideas. The language is different, the notation is different, but everyone has more or less the same math.

All of it? Arithmetic, geometry, sure. But all the same math? All this stuff you're saying about symmetry groups, Four-Color Theorems, infinity versus the continuum. You're saying every culture has its own translations of all these exact ideas? That's hard to believe.

Okay, the short answer is no.

Aha!

Because each culture picks different areas of math to focus on! The Mayans got really deep into calendars, the Pythagoreans were obsessed with ratios. So those are the areas they end up developing.

For sure that has to do with different values and priorities, aesthetics, cultural things like that. That doesn't make the math itself any less valid! Whenever you have multiple cultures looking into the same math, they always find the same thing.

Always?

As far as I know.

Hm. So which culture is this? You're pulling from
some particular math canon, right?

How do you mean?

When you say the three main areas of math are
topology, analysis, and algebra, that's a reflection of
a particular culture, no? And when you're telling
me what's been proven or not, that's according to a
particular community of peer reviewers.

True, true. I guess it feels like now is a bit different
than historically, in terms of math cultures. Now
we have globalization, with airplanes and the
internet and everything. When you talk about
"math" in any major world city, or if you study
math at any big-name university, you can be in
Nairobi or Shanghai or Cambridge, you're going to
learn about the same stuff.

Still, that's a tradition. That modern, global math
tradition—where did it come from?

Well, it was imposed on the rest of the world by
Europe, through colonization and imperialism. But
the actual math itself? In terms of the notations and
what subjects we focus on, and the specific meth-
ods we use? If you look into it it's pretty much
all from the math tradition of Arab and African
Muslims.

Right, I thought so! I mean, numbers
are literally called Arabic numerals.

Exactly. And "algorithm," that's someone's name, Muhammad al-Khwarizmi. It's like "Phillips-head screwdriver"—algorithm just means "This was al-Khwarizmi's idea." Algebra too, that's a mispronunciation of an Arabic word, *al-jabr*, which there wasn't a word for in any European language. It was used to describe when you flip a term to the opposite side of an equation. You know, *algebra*.

So that's all from Africa, yeah?

Somewhere around what we now call North Africa and the Middle East. It wasn't split up back then. It was just a network of societies that traded and exchanged ideas with each other. And for half a millennium, while Europe was busy fighting off the Vikings and each other, the Muslim world had a long stretch of peace and prosperity. Lots of time to kick back and think about math!

That's when they came up with most of the techniques of arithmetic and algebra we all learn in school. You know, solving for unknowns, decimal points, irrational numbers, polynomials and quadratic equations and completing the square, all that. When you think about the focus we have today, the global math culture, it's all about abstraction and manipulating symbols and organized, algorithmic procedures. That's rooted in the Muslim tradition.

But to be fair, no culture ever did math in isolation! The "Arabic number" system was borrowed from

Hindu scholars, who wrote their math in Sanskrit poems. And China had the abacus, you know, everyone found a way to do it. "Al-jabr" is just the one that caught on in Europe, right next door. For hundreds of years they used translated Arabic textbooks to teach math at the best European schools.

> That's neat, that's good to know where it all comes from. But I still feel like you're dodging a bit.

Dodging?

> In terms of the modern global math culture. So the ideas come from the Islamic world. But as you said, it was Europe that made it global. And correct me if I'm wrong, but it seems like when you're talking about modern math—not the classic stuff like solving for x, but these crazy theorems they only teach in college—it seems like most of the people you hear about are European, right? Which is definitely suspect in terms of me believing this is some universal, objectively true thing not rooted in culture.

Well look, I'm not a historian, but you and I both know the last few centuries have been a violent and oppressive time for people in colonized countries—which is basically everywhere outside Europe. A lot of these theorems were proven at a time when most of the world wasn't allowed anywhere near the ivory towers of academia.

> Right! Besides, if your whole society is getting stamped out and reorganized, your top priority

probably isn't going to be sitting around with chalk wondering what's the deal with shapes.

I suppose. It's really unfortunate how these arbitrary historical events can deprive us of so much potential math talent. I always think, when I read a biography of a famous male mathematician, where would this story have gone differently if he'd been raised as a girl in that time?

That's a depressing thought. Clearly math is a powerful thing, so it's not surprising that people try to hoard it and keep it exclusive.

Such a weird urge. The whole point of math is that it's supposed to be completely universal!

Right, okay. So say you're right that math is this perfectly true thing, and it being dominated by white boys is just a recent historical accident, having to do with politics.

Right.

Then doesn't that still create a bias? I mean, if it's mostly white guys in the room, reviewing the papers, grading the tests, won't that affect what gets taught and what gets accepted as true?

I still don't see how it changes the underlying math . . .

Really? How could it not?

I'll give you this: I can see how it could affect the specific things that get studied and prioritized, and it might even affect what types of ideas people come up with or miss out on. But the math itself was already there. I think if you prove something is true, it really is true.

Hm.

Alright, you can tell me about this thing between true and false.

Okay, great—I think you'll actually like it. It supports your point that math might all be made up.

I'm listening.

You should know, though, this result was a real big deal for mathematicians at the time. It shattered the vision of math as a perfect, pristine crystal of truths and falsehoods. It added this murkiness that none of the higher-ups in math wanted to acknowledge. We still haven't fully recovered.

Alright, what is it? What is there
other than true and false?

Hold on, I want to give some context first.
Need to set the scene.

Fine, sure.

This is all happening about a hundred years ago.
Height of imperialism, leading up to the world
wars. Most of the characters involved are probably
what you'd expect: a bunch of rich white men, people who grew up with expensive tutors and lots of
free time. Some royals and earls thrown into
the mix.

Sounds about right.

Now there was a minor panic going on in math
around then. It's along the lines of what you're saying: How do we know any of this is true? This was
around the time when abstract algebra was really
blowing up, all this research into deep structure
and the nature of logic itself. A lot of math was
being reduced to axioms, formal systems, dots and
lines, moving symbols around according to arcane
rules. And people started thinking, what is going
on here?

That makes sense. As you go from basic intuitive
arguments to these formal games of abstraction,
you'd start to get less confident that what you're
doing is legit.

Right, it gets spooky. Why does this work?

It's cool that these guys would
admit they had concerns.

Well, not a lot of them did, but it only takes one or
two to cause trouble. There was this one Dutch
topologist who started coming out and saying
things like "Math is an extension of human intuition" and all sorts of embarrassing philosophical
claims which threatened the legitimacy and esteem
of formal math.

And the other mathematicians were furious! Some
of them arranged to get this topologist booted from
the board of *Mathematische Annalen*, which was a
top math journal. They didn't want him influenc-
ing other people and getting these blasphemous
ideas out.

Doesn't that kind of undermine the legitimacy
right there? If math can be influenced
by petty politics like that.

Right, well that was never intended to be the final
word on the matter. It was a temporary fix to
buy some time. What these guys ultimately wanted
was to *prove*, once and for all, that mathematical
proof is the ultimate decider of what's true and
what's false.

So they wanted to prove that math
is legit, using what? Math?

I know, in retrospect it's pretty wild they thought that would work.

> Wasn't that obvious? I feel like they all must have immediately seen the problem with that.

Well, it's not so simple. They weren't trying to prove that math is "legit"—that doesn't really mean anything. People use math all the time and it always seems to work, so it's already pretty legit in that sense.

What they wanted to do was build a sturdy foundation for mathematics, a rock-solid ground floor that everything else could rely on. Until then, the concept of "proof" relied on intuition: Is it convincing? That was starting to feel flimsy and fallible, especially when you're dealing with weird, abstract objects. So they wanted to switch over to a new, rigorous form of proof, something organized and systematic that wouldn't depend on who's doing the proving.

> They wanted to take intuition and subjectivity out of the picture, is what you're saying.

You could put it that way, sure.

> I don't see how that's possible. You can make up a new kind of proof with a bunch of rigid rules, but everyone still has to agree on what those rules are. It's not like they fell out of the sky—people made them up based on intuition and subjectivity.

Okay, but here's the catch. The idea was to work
their way up, starting from basic logic.

Hm. Explain.

So yes, people can have principled disagreements
about what should qualify as a formal proof.
Maybe you think these newfangled computer
proofs are unreliable. Or maybe you think we
shouldn't be messing around with infinity, we
don't really know what we're talking about, and so
any proof involving infinite sets isn't trustworthy.

Yeah, there's a lot of room for disagreement.

Totally. It's been argued that irrational numbers
don't really exist. Some people even think *fractions*
are a little suspicious, and we should stick to whole
numbers only!

That's hilarious. That's actually really interesting,
I'd love to talk to someone who thinks that.

But the idea is: The lower you go, the sturdier it
gets. We're pretty confident that basic counting is
legitimate, right?

Though I'm sure someone disagrees with that, even.

Actually, yeah, there's a mathematician who
argued that even *whole numbers* only go up so
high—that really, really big numbers don't exist.
But no one thinks that idea holds much water.

So that's the plan, then. These big-name mathematicians, they were going to bootstrap their way up. Starting from the absolute basics, what they call zeroth-order logic, they'd prove all of first-order logic, then elementary arithmetic, then they'd use that to prove things about irrational numbers, and then imaginary numbers, and one by one they'd prove every known mathematical truth within this one sturdy system.

And then all the doubters and the haters would have to write a very nice formal apology letter.

> They wanted to re-prove every single theorem in terms of basic logic?

It's not as bad as it sounds. You can sort of "swallow" a higher field into a lower one. You find a way to take any proof in the higher field and translate it down into a proof in the lower field, using simpler objects and rules. And so on down to basic logic.

> Okay, makes sense. What if someone doesn't believe in basic logic?

Seriously? You don't even think *logic* is objectively true? "If P is false, then 'not P' is true"—you deny that?

> No, not me! I believe in logic. And I can let you assume basic logic, since it sounds like you're heading in a direction that's going to support my side anyway.

But that's exactly the point: You still have to assume something! You can't prove anything just out of thin air. You have to start somewhere, with some first premise, and that came from your intuition.

I mean, after a certain point, can't we just say the ground floor is our basic sanity? "*A* implies *B*, and *A*, therefore, *B*." No?

It's still an assumption.

Fine. You're right, you can't prove anything to someone who's being obstinate. If you're not on board with basic logic, you're not going to get on board with the rest of this whole program.

But that's your loss! Look what you're missing out on! If this bootstrapping project succeeds, we've just placed every true mathematical fact into one consistent framework, one neatly organized structure.

Fair enough, because that's a worthwhile goal on its own.

Right, wouldn't that be compelling? A densely packed lattice of knowledge that contains every true statement!

"The tree of knowledge of True and False."

Yes, exactly. And you're going to turn that down, what, because you don't believe in logic? Come *on*.

Fine. I can see that. When you all agree on some
principles of basic logic, you get this big shared system of math knowledge.

And math is a foundation for physics, which is a
foundation for chemistry and biology, which are
the foundation of human behavior, and so on and
so on. We might be able to bootstrap our way up
from basic logic to *every* subject, gather *every* true
fact into one single tree. And then we could finally
achieve objectivity about everything—where objectivity isn't some complex, hazy thing anymore, it's
defined as "exactly what's in this mathematical tree
of truths."

That's the idea, at least.

I can see how that'd be an addictive idea, especially
for a bunch of earls who wanted to feel like they
were right about everything.

Right, so, these guys, these royals and scholars,
they get to bootstrapping. And they're doing a
pretty good job. They figure out how to put real
numbers in terms of integers, and they get all integers out of just the number zero and the idea of
"plus one."

That's pretty cool.

They're really pulling it all together. They get to the
point where they're almost done. There's just one
step left.

Wow, one step? So they really almost got calculus
and everything in terms of just basic logic?

Yeah, well, they have a lot of time on their hands.

What's the last step?

They have to prove that arithmetic is complete.
Their version, the little rinky-dink version they built
out of zero and "plus one." They have to prove that
it's good enough to prove all truths of arithmetic.

Okay. Not sure how you'd prove
something like that, but okay,
that was the one step left.

And they're getting really excited, they have the
champagne bottles ready to go. They really think
they're about to do it! All of math, out of just six
axioms and four inference rules. It was a cultural
phenomenon in those circles. They wrote books
called things like *Principia Mathematica*. And of
course there were people who called them crazy,
said they could never do it, that it was all mean-
ingless. But no one listened to them, because they
weren't on the board of *Mathematische Annalen*.

So what went wrong?

It was a disaster. Humiliating. The blow was dealt
by one of their own.

Drama!

A fellow named Gödel, rhymes with turtle. He was the same guy who proved that their version of first-order logic was complete. Big hero! He was in his twenties when he proved it, which is plenty of time for another big breakthrough. Seemed like he might be the one to show arithmetic is complete too.

> Let me guess: He proved their little model of arithmetic wasn't complete.

So much worse.

> What did he do?

He proved that *every possible* model of arithmetic is incomplete.

> So . . .

So the bootstrapping project is impossible. You can't prove all the truths of math in one formal system. You can't even prove all the truths of arithmetic in one formal system.

> Wow. How do you prove that?

It's the same sort of argument you use to prove the continuum can't be put in a list. You take any system that supposedly contains all truths of arithmetic, and you find a truth it's missing. You basically find a sentence that says, "This statement can't be proven from the axioms."

Huh . . . Okay, neat, I could imagine
how that would go.

And if they try to add the missing truth in as a new
axiom, well, you can just run the same process
again and find a new sentence that says, "This
statement can't be proven from *those* axioms."

Nice. And the important thing is that everyone
else agreed his proof was valid.

Oh yes, they did. No one could deny it. So much at
stake, and no one could find a flaw in Gödel's logic.
It was airtight. They had to publish.

Wow, I respect that.

That was that. The dream was shattered. They had
to put *Principia Mathematica* far away where no one
would ever find it. Some of them quit math and
went into philosophy. Some of them worked on formal semantics, linguistics, theory of computation,
stuff that later turned into early programming
languages.

Not surprising—those axiom systems sound similar to coding languages. All about *ifs* and *thens*, lots
of variables, strict rules.

And algorithms too. They had already worked out
step-by-step algorithms for how they would use
this perfect, bootstrapped system to automatically
generate new truths. Early computers, they weren't

a place to look at pictures of each other. They were
designed to carry out systematic computations like
that. You know, to *compute*.

> Alright, very nice. Nice story. They almost built an
> automatic truth machine, and then they didn't,
> because that's impossible.

> So what's the status between true and false?

Well, I shouldn't say "there's something between
true and false" so matter-of-factly like that. People
disagree endlessly about what Gödel's proof means,
how we ought to interpret it. What do you think
it means?

> What it means?

The idea that no formal system of proof can prove
all mathematical truths.

> Hm.

> I guess I'm not shocked. You can talk about univer-
> sal truths and objective proof, and maybe that stuff
> exists—who knows, it might! But as a practical
> matter, *in practice*, "proof" always refers to what-
> ever people find convincing. And that's based on
> intuition and subjectivity and social context, there's
> no way around it.

> There have always been groups of people who
> believed they were right, right? Not just the way

everyone thinks they're right, but really *Right*, you know, objectively, the kind of thing where God would agree with you. And they've gone to great lengths trying to prove it's not just in their heads and everyone who disagrees is mistaken, and they've usually ended up looking pretty foolish.

So these people tried to take the intuition out of math, reduce truth to a formula. It's audacious, I'll give you that. It sounds like they had all the resources they could possibly want to give it their best shot. So it didn't pan out. To me, that means truth is a slippery thing and it doesn't conform to human notions of order and control.

That's a perfectly valid view, I see where you're coming from.

And you? How do you interpret Gödel?

Well, I go back and forth.

Maybe I'm old-fashioned, but I do still think math is true! And I do think math teaches us a lot about what truth is, and what kind of structure or rhythm it has. I think proof is an important thing, logic is an important thing, these aren't just foolish attempts to constrain reality to our will. I think they actually reflect something about how the universe fits together.

In math, some things are provably true and some things are provably false. And then, according to

Gödel, some things are neither. Some things are provably unprovable. "Independent from the axioms of ZFC," as they say. Questions without answers, and not because we haven't found them yet—the truth value is simply undefined.

That leaves us with two options, both bad. We can say that there really is a third category: *unknowable*, maybe, or indeterminate. Or we have to accept that truth is not identical with provability, that there are true statements that can never be proven, and our only means to access them is wishy-washy things like "metaphysical intuition."

But that's just my view, and we can agree to disagree.

Well, based on that, it actually sounds like we agree.

We definitely don't.

some philosophies of math

platonism—mathematical objects
really exist in some "platonic realm"

intuitionism—math is an extension
of human intuition and reasoning

logicism—math is an extension of
logic, which is objective and universal

empiricism—math is just like science:
it must be tested to be believed

formalism—math is a game of
symbolic manipulation with no
deeper meaning

conventionalism—math is the set
of agreed-upon truths within the
math community

a logic riddle

Three very logical people are standing in a line so they can only see who's in front of them.

A hat seller shows them three white and two black hats. She places a hat on each person and hides the remaining two.

She asks, "Does anyone know what color hat they have?"

No answer.

"Now does anyone know what color hat they have?"

No answer.

"Now does anyone know what color hat they have?"

One person answers.
Which person, and what color hat?

a harder logic riddle

Three identical triplets guard
three identical doors.

The youngest triplet always lies. The
oldest always tells the truth. The third
triplet is the trickster and answers
as she chooses.

Behind the trickster's door is certain
doom. The other doors are exits.

You may ask one question to one
triplet (without knowing which)
and then you must choose your door.

What do you do?

Modeling

models

automata

science

∫ models

*O*kay, okay, I hear you. What's the point of any of this, right? Axioms, double and triple tori, continuum-sums, wallpaper symmetries. For what? In the pointed phrasing of math students around the world and throughout history:

When will I use this in real life?

I've tried to avoid addressing this question directly because (and I promise this is the last time I'll remind you of this) professional mathematicians really don't care about real-world applications. That's the domain of *applied* math, the opposite of *pure* math, which should give you a sense for how the word "applied" is meant to sound. But here we are, with a good chunk of pages to go, having already run through the three main branches of pure math, plus a little history and philosophy. So I'll entertain the question and say a thing or two about applied math, even if it'll get me in trouble with the hardcore types who find this "real life" stuff irrelevant and distracting.

In particular, this last section is about modeling. Modeling is how math connects to the real world. Of course there are lots of different ways math turns up in the real world, but modeling is a sort of general framework that lets us see all these connections clearly. It gives us a convenient way to talk about the connections, so we can explore them and learn new things.

A model consists of two main ingredients. There's the way the model itself works: a set of internal, mathematical rules that determine how everything inside the abstract model-world operates. And then (this is the important part) there's some kind of translation process that connects the model back to the outside world.

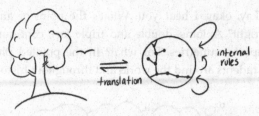

Of course I've skimmed over all the gritty details, but even from this rough description you can see what an arrangement like this would allow us to do. We could observe something in the real world, translate it into the language of the model, follow the internal laws of the model to infer new truths, and then translate it back into our reality. We could, in other words, learn things about the real world by taking a detour through a fictional, mathematical world. This is new.

Let's look at an example: music theory. Music theory is an abstract model of how music works. You take real-world music, a complex and chaotic parade of vibrations in the air, and you translate it into a symbolic system of notes and chords. Inside that abstract system, there are certain rules or guidelines (for a given genre or musical tradition) about which notes work with which chords, which sequences of notes will sound tense or sad or funky, and which chords typically follow which other chords. These are all the makings of a model. We have a simplified representation of a real-world thing which makes it easier to manage, analyze, and predict that real-world thing.

Yes, we lost detail when we abstracted. It's not a perfect translation, and the model-world isn't going to be isomorphic to the real world. That's fine. If you're playing in a jam session, you mostly just need to know the chord progression, the rhythm, and what key you're in. If you tried to analyze every aspect of the audio flow coming into your ears, you'd get hopelessly lost. Instead, you strip it down to the basics—you abstract. "Notes" and "chords" aren't tangible, real-world entities. These concepts live in model-world, they have internal rules of engagement, and they correspond back to sounds in the real world. They're useful theoretical constructs.

This is the key to a good model: a smart stripping-down process that takes us to a basic but still useful unit, like a note or a chord. When we're working inside the model, we temporarily pretend these things really are unbreakable atoms with fixed laws of behavior. This isn't strictly true: A note is actually a mishmash of overtones, echoes, and reverb all bouncing around, pushing up against your eardrums. But if it's useful to build a tiny model of the world where it *is* true, where a note is just a note, well, what's the harm in that?

Sometimes this stripping-down process goes a little too far, and we do have to be careful about drawing real-world conclusions from oversimplified models. It's often convenient to make assumptions that are not quite true, or even ones that are demonstrably, laughably false. We just have to strike a good balance between simplicity and usefulness. There's an old joke about an academic who's called to a dairy farm to help increase milk production and says, "I have a solution. We assume a spherical cow. . . ."

Here's another modeling example, from economics. Say there's some product that lots of people want to buy. Hot sauce, for instance. And then something happens, like a pest infestation in the chili fields, that reduces the amount of hot sauce being produced. What happens next is predictable: The price of hot sauce will go up. This is the kind of real-world regularity that lends itself perfectly to modeling. When there's a sudden shortage of something, its price typically rises.

Of course, a "price" isn't really ever just a single number. It depends on where you buy your hot sauce, who's selling it to you, how that person's business model works, maybe even how wealthy that person thinks you are. When the shortage happens, sellers who don't hear about it immediately might keep selling hot sauce at the original price until it runs out. Or buyers who don't know about the shortage might refuse to buy it at a higher price. Or within a certain community there may be an expectation of what the "fair price" of hot sauce is, and sellers could be ostracized for jacking up prices. It's hard to imagine something more complicated, with more moving parts under the hood, than a price.

But when we're modeling, we can make the simplifying assumption that price is just a single number, the same every-where. We can also assume that the "demand curve" and "sup-ply curve" (more abstractions invented for our modeling convenience) are simple functions that, depending on price, tell you exactly how much hot sauce will be desired and how much will be produced. We can assume that, in a "competitive mar-ket" (another abstraction), everything will settle into an "equi-librium price" (and another). Within the theoretical world built from these assumptions, we can solve for the equilibrium and convert it back to a prediction of what the real-world price will be. And in some cases, this supply–demand model actually makes pretty decent predictions.

Of course, we have to be careful about which assumptions we make. One of the standard assumptions of neoclassical eco-nomics is that humans are rational actors: that we have innate and consistent preferences, that we seek out the highest-paying jobs and the lowest-priced products, that we have complete information about pretty much everything. Most of this is not, in the real world, accurate. These are simplifying assumptions that let us make predictions. If the predictions tend to come true, great! The model is useful. That doesn't mean the assump-tions are true. There are plenty of ways in which humans very much don't act rationally: We're overly risk-averse, we don't plan for the future well, we buy expensive things to flex our wealth, we discriminate, we give jobs to friends and family over more qualified outsiders, the list goes on and on. If you try to apply the standard models in these cases, they'll break down and make poor predictions.

This is a crucial point about modeling in general: A model works only within a certain scope. The assumptions you use to make good predictions in one area (like economics) could be totally different from the assumptions you use to make good predictions in another area (like sociology). This doesn't mean one model is right and the other is wrong. It just means you have to know when to use which. If you think you have a single, consistent model that works in all contexts, you're probably just ignoring or downplaying the contexts where it doesn't work. No model is sacred.

One more example: Have you ever watched a movie and, about halfway through, been able to predict most of what happens for the rest of the movie? When you think about it, that's a pretty remarkable feat. How can you see the future like that? You must have a mental model of "how movies usually work" that you've developed from watching movies all your life. You simplify the flow of information coming into your ears and eyes, turning pixels into abstract units like characters, dialog, motives, relationships. Then you apply some unspoken rules: "If they show a loaded gun, it'll get shot before the movie's over" or "That character who's super racist will definitely get their come-uppance" or "Around the last twenty minutes of the movie they'll break up over this character flaw but then he'll learn his lesson and make a grand romantic gesture and they'll reunite dramatically and live happily ever after." Sure, these aren't strict mathematical rules, and the predictions might not be accurate every time, but you're still doing some rudimentary modeling. You're building a set of rules in your head that you can apply across a variety of similar real-world circumstances.

And really, when it comes down to it, this is what's going on in our heads all the time. We interpret the world around us not as flashes of light and sound; we chunk it into things, entities, units of analysis we expect to behave in certain ways. We see something we categorize as a "car" and something we categorize as a "green light" and we think, *Cars typically keep driving through green lights. If I cross the street now I'll likely be hit.* Human perception and cognition are all about pattern recognition, and to recognize patterns we first have to abstract the continuous, fuzzy reality around us into discrete objects that can behave in patterned ways.

Notice, also: Models don't have to be mathematical. The internal rules of model-world can be rough and qualitative, things like "Opposites attract" or "Birds of a feather flock together." If anything, it should be far easier to build these kinds of non-mathematical models. A model that makes precise numerical predictions, after all, is very easy to prove wrong.

Which is why it's surprising that our world makes itself so intelligible to mathematical modeling. A remarkable number of things, if you pay close attention, are practically screaming for us to use math to describe their behavior.

Here, take any small object. Your keys will do. Toss it up from your left hand and catch it in your right hand. The path it makes through the air is a perfect parabola. No matter how you throw it, it'll always follow a parabolic path. It recreates a mathematical object, a precise geometric shape, in real life!

Or take a piece of string and dangle it between two points. It'll settle into a shape called a catenary, a perfect replica of a graph called the hyperbolic cosine. Telephone wires, unweighted necklaces, velvet VIP ropes—no matter the material, it'll always make this same shape. (The formula for this shape, by the way, involves an irrational number called *e* that arises from the study of compound interest, and which has absolutely no right being in the equation for how strings hang.)

One more shape. This one's a bit more involved. Set up a camera on a tripod and point it at the sky. Pick a time of day to take a picture. Leave it in the exact same position, and take a picture at the same time the next day, and the next, and keep doing this every day for a year. The path of the sun over the course of a year will trace out a mathematical shape called an analemma.

I'm giving examples of complicated shapes, because simple mathematical shapes are so commonplace in nature that we hardly notice them. When you blow soap bubbles they form perfect spheres. Drop a pebble in a pond and the ripples will travel out in perfect circles. These examples don't seem quite so surprising, but they also point to there being some sort of mathematical logic operating behind the scenes.

This bizarre recurrence of mathematical phenomena in the natural world goes far beyond physical shapes. Another familiar example, which we really shouldn't take for granted, is the bell curve: a formula for predicting the distribution of almost any numerical property in any naturally occurring data set. Here, for instance, is the distribution of women's height in the United States:

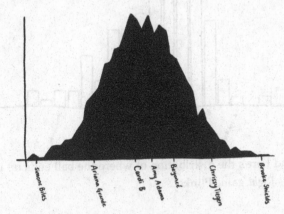

And here's the distribution of scores on the multistate bar exam:

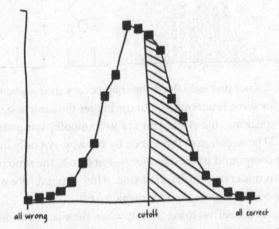

Here's the distribution of song length for all Billboard number one hits of the aughts:

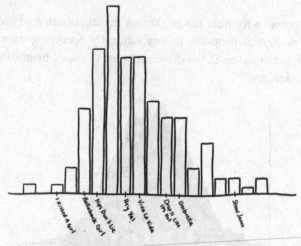

And here's the distribution of where the ball ends up in *The Price Is Right* game Plinko:

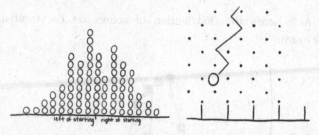

No, it's not precisely the same shape every time—you have to allow for some randomness. But the bigger the sample size, generally speaking, the closer you get to a smooth, symmetric bell curve. (The equation of this curve, by the way, not only includes *e*—the compound interest number—but also π, the ratio of a circle's circumference to its diameter. After a point, doesn't this start to feel like some kind of cosmic joke?)

This is the eeriest thing for me, when the exact same formula pops up in different fields of study, in entirely unrelated contexts that don't seem like they should be analogous. So, for instance, the famous gravity equation tells us the force of attraction between two macroscopic objects if we know their masses:

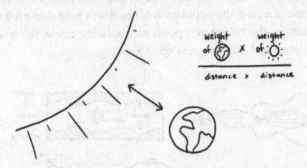

But it also tells us the force of attraction or repulsion between two microscopic particles if we know their charges:

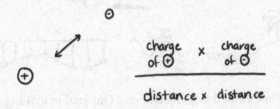

And (brace yourself) it also gives us a good estimate for the amount of trade between two countries if we know their GDPs:

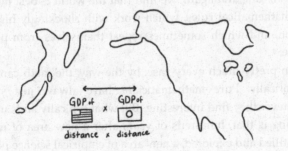

Better yet, the mathematical process known as "simple harmonic motion" identically describes the vibration of a plucked string, the length of a day* and average temperature over the course of a year, the population of species in predator–prey relationships, the height of a point on a rotating circle, the level of the tides, and the compression of a spring.

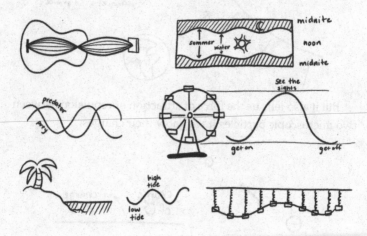

What the hell is going on here? Our goal in making models, remember, is just to be useful, to find a nice and convenient system to summarize what we observe in an orderly way. The rules of a model can take any form, rough or precise. But for some reason, time and again, we find that the world is best modeled by mathematical rules, which work with shockingly high precision, and which sometimes repeat themselves from place to place.

In pretty much every case, by the way, the math came first historically. Pure mathematicians have always just studied whatever they find interesting. But what typically ends up happening is that, hundreds of years after a new area of math is identified and explored, a new area of empirical science pops up that requires exactly those same mathematical concepts and results. We're not inventing math to fit our world—we're discovering what math is out there, and then later realizing that our world happens to look exactly like it.

How can we explain this? Why is the world so susceptible to mathematical modeling?

The most honest answer is that no one really knows for sure. This is a hot topic for debate among philosophers of math, and I'm not going to pretend I know the answer. Within the pure math community, though, there's one theory that seems to be very popular. People won't come out and say it quite like this, but I've run it by enough people to feel confident saying a lot of us believe it's true.

Maybe we observe mathematical patterns in nature because *the world itself is made of math*. Maybe the universe is fundamentally mathematical in character, and there's a One True Model that perfectly describes its behavior.

Let's not mince words: That sounds insane. But hear us out.

automata

*H*ow could a world be made of math? Let me at least make the idea plausible.

You've likely seen worlds made out of math before. They're called simulations. Sure, most of the time a simulation is just a dinky little world with not too much going on, a couple objects with predictable behavior acting out some prescribed scenario. That's a long way from our unpredictable, intricately detailed world, but we have to start somewhere.

Mathematicians invented simulations long before there were any computers to play them out. If a simulation is simple enough, you can do it out by hand—it's just a paper-and-pencil game to pass the time. We usually say "automaton" instead of "simulation," but it's the same idea. There are predefined rules for how everything moves, you pick a starting setup, and then you let it run and see what happens.

Here's a simple simulation we can carry out by hand. The world is a one-lane road split into discrete boxes.

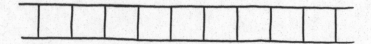

The only object is a car. The car behaves according to the following rules of motion:

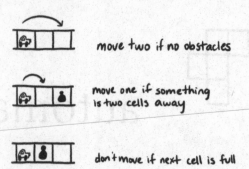

move two if no obstacles

move one if something is two cells away

don't move if next cell is full

If we put a single car on the road and "press play," it's not hard to guess what'll happen next.

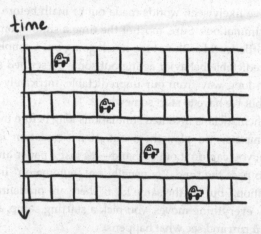

time

What if you start with a lineup of five cars? It's a little more work, but not too much. You go car by car, checking how far it should move, and repeat for each time-step. You're playing the role of the computer, aka you're doing some rudimentary computing.

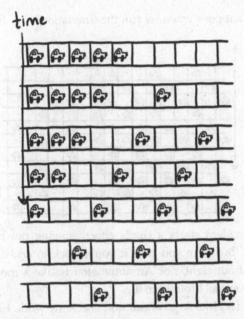

This is a very basic automaton (one-dimensional, discrete, deterministic), but even this is good enough to recreate some real-world phenomena. For instance, we can add a rule for rubber-necking:

move one if passing a
gawk spot

Now let's start with an infinite line of cars, spaced two car-lengths apart, approaching a gawk spot.

What happens when we run the simulation?

The accident starts a ripple effect, slowing down the cars behind it, but once you pass it you're back to cruising speed. Sounds about right, no? An automaton is like a model set in motion, a model brought to life.

If you feel like it, go ahead and check my work: Each car, at each time-step, moves according to the four rules of motion. You can also rerun the simulation yourself on a piece of graph paper, or try out new starting scenarios to see what happens. Some people find this sort of thing monotonous and painful, others find it entertaining and therapeutic.

This is still in the category of "way too simple to be anything like the real world." Yes, we can recreate some basic patterns, but we all know real human drivers are way more complicated than four formulaic rules. They have distractions and muscle spasms and places to be. Besides, if we're aiming for a genuine Model of Everything, we have to recreate not just the cars moving down the road, but the birds flying by, the churning of the engines, the state of international affairs, the pumping of a blood vessel in the left palm of an acrobat napping a few towns over. A basic automaton like this car example is clearly not going to measure up.

Fair enough! We're just getting warmed up. Now let's look at the most famous automaton of all time. It's still way, way simpler than our real world, but it generates some behavior that might make it plausible that our world could be a very complex automaton.

It's called, appropriately, the Game of Life.

Like the car example, this is a world composed of discrete square cells. In the Game of Life the world is a two-dimensional grid, extending infinitely far in all directions. Each cell has two possible states: on or off. Unlike the car example, these cells don't represent any particular real-world thing. They're just squares that can be on or off, black or white, filled in or empty.

There are three rules that determine how everything plays out in the Game of Life. Each cell decides whether to be on or off in the next time-step by checking how its eight neighbors (including diagonals) are doing in this time-step.

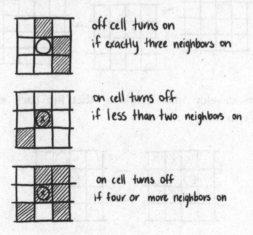

off cell turns on
if exactly three neighbors on

on cell turns off
if less than two neighbors on

on cell turns off
if four or more neighbors on

This automaton is a little harder to carry out by hand, since there are more cells to check each step. But if you're organized about it, and you come up with a consistent notation, you can run the simulation from any starting setup and see what happens.

Some setups settle into a stable state and stay that way forever.

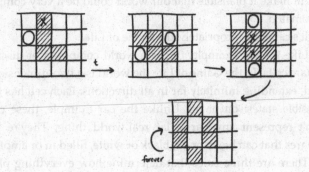

Others quickly fade into nothingness.

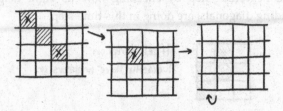

Others develop into "blinkers" that flip back and forth between two states forever.

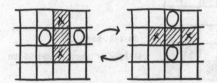

Some setups are "gliders" that cycle back to the same initial pattern, but shifted down and to the right.

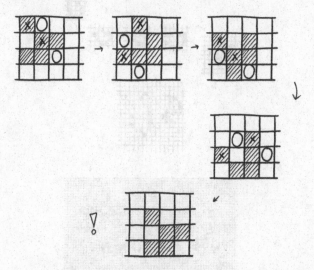

These are called gliders because, over the course of multiple cycles, they glide indefinitely across the stage.

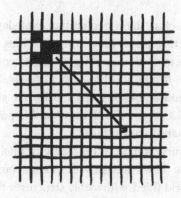

And then some other setups, well . . .

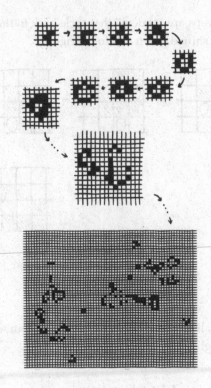

This five-cell setup, the R-pentomino, explodes into an eco-system of interacting parts, generating still lifes and blinkers, shooting off gliders, evolving and growing to cover a massive area of the stage. It finally settles down to a stable, repeating pattern after a thousand time-steps, but for the stretch of time before that it looks very, well, lifelike. (At this point, doing the computation by hand is not advised.)

This isn't too uncommon in the Game of Life. Sometimes a reasonably simple initial pattern will spontaneously generate a large, chaotic world filled with stable structures that move and interact over time in interesting, nonobvious ways. Sound familiar?

There's a starting pattern that leads to infinite growth, by shooting out an endless stream of gliders at regular intervals. There's a pattern called "Sir Robin" that drifts across the stage like a knight in chess. There's one called "Gemini" that computationally generates an exact replica of itself, after millions of time-steps. (Of course: There's a ferociously dedicated online community that hunts for these things and gives out Pattern of the Year awards.) Just about any kind of behavior you could imagine taking place on a grid of black-and-white pixels, there's a pattern for that.

Still think it's too simple to be our world? You're probably not wrong. We don't live in a flat, discrete, black-and-white world. This particular Game of Life—this stage, these rules—are arbitrary, chosen not because they reflect reality but because they're easy to work with. But we can invent automata with whatever rules we want.

We can invent an automaton on a hexagonal stage:

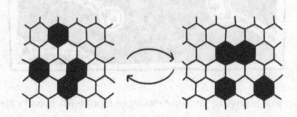

We can make one with more than two different cell states:

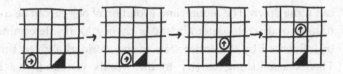

Depending on what rules we choose, these fictional worlds can exhibit vastly different sorts of behavior. Some worlds quickly collapse to nothing, no matter what pattern you start on. Others explode out like the Big Bang, from a single pixel.

If you don't like it being made of pixels at all, no problem: We can make automata that take place on a continuous stage. Here, the rules of the game aren't about "number of neighbors turned on" but about "percent of local environment turned on." Here's an automaton called SmoothLife, which looks creepily like something going on in a petri dish:

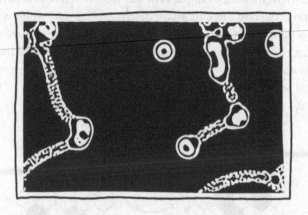

I'm just showing you one example from each broad category of automaton, but keep in mind that the options are limitless. After you've picked a dimension and space to set your world in, and after you've picked a set of fundamental objects or cell states, there's still an infinite bouquet of possible rule-sets you can write down. The motion and evolution of objects can be continuous or discrete, deterministic or chancy, locally determined or influenced by the entire state of the world at a given time. There's astonishing variety to the worlds you find, just by slightly changing a single parameter in a single rule.

Here, for instance, is another continuous automaton, called Crystalline:

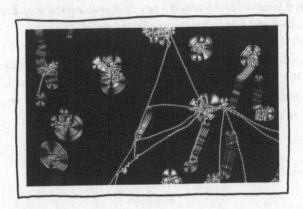

Just kidding—this one isn't an automaton, it's a real-life photograph of liquid crystals under a microscope.

So is it really that hard to imagine there's an automaton out there that generates, well, *this*?

If this makes you feel existentially queasy, consider this paragraph a real-life spoiler alert. The last chapter of this book introduces a special automaton, called the Standard Model of Particle Physics. It's a continuous, three-dimensional automaton with seventeen fundamental objects and about twelve rules of evolution. From certain starting conditions, when you press play, well, it's pretty eerie what happens next.

The Standard Model is the best model we've found so far to recreate our world in purely mathematical terms. It's not perfect, but it's close enough to be legitimately creepy, like a weird dream-version of reality. Or, depending on your religion, it might feel like a new, higher level of reality that makes everyday life look like a weird dream.

If you don't want to see this (and it's perfectly legitimate to not want to peek at the source code), then I suggest you put this book down now. Really, I won't be offended! I hope you've enjoyed yourself and learned a few things along the way. End of book. Bye, and have a great rest of your week!

But if you *do* want to see it, if you want to zoom in so far you can see the pixels, keep reading. This last chapter is for you. But consider yourself warned: All I'm offering is, well, not even the *truth*, necessarily, but one unreasonably useful way of looking at things.

science

*H*ere are the rules of the mathematical game we call the Standard Model. The rules aren't set entirely in stone, and in fact we know that the model we're working with at the moment isn't exactly right. But it's pretty close, and here are the rules of the game.

Start with an empty three-dimensional space. Which one? We're not sure—remember, topologists have a whole catalog of three-dimensional spaces which look locally like . . . *this*. Cosmologists have done some work to figure out the shape of the universe, based on their own set of models and assumptions. It doesn't really matter for our purposes. Let's just say we're working with the basic, infinite, non-curved three-space.

So you have a large empty space. At any point in the space you can place an infinitesimally small point-object called a "particle." The space is continuous, so I really mean at *any* point. No square cells here. And these things we're calling particles, don't think of them as tiny, shiny orbs. They're literally just points.

They take up no space. They are mathematical points with zero size.

Not all particles are created equal: They have slightly different properties which determine how they'll move. Whenever you create a particle, you have to give it a "mass" (a positive number) and a "charge" (positive, negative, or zero). And you can't just pick any mass and charge—there are only seventeen legal combinations of mass and charge to choose from. We call these combinations the seventeen fundamental particles, and we give each one a cute name like "charm quark" or "tau lepton."

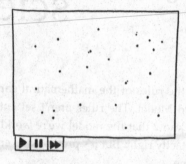

When you press play, what happens to the particles? They move through the space and interact. Like any automaton, there are precise computational rules which tell you what each particle will do next. Generally, they move very quickly and in straight lines. The only exceptions are interactions: when a particle decays, or when two particles come really close together. Then we have to consult our handy lookup table of interactions to find out what happens next. Depending on the identities of the interacting particles, they might collide and scatter off in different directions, or combine to form a single particle, or (if they come at each other fast enough) they may spew out a jet of new particles.

If you're curious, here's a list of all fundamental particle interactions under the Standard Model:

The first one shows, for instance, an electron absorbing a photon and changing direction. These interactions can also run in reverse, e.g., an electron can also emit a photon and change direction.

I'm being vague on the details, but I don't have much choice. This isn't the Game of Life, where you just count up boxes to find out what happens next. The exact rules for particle interactions in the Standard Model are absurd, to be honest with you. The calculations involve continuum-sums and imaginary numbers and coupling constants and all sorts of ridiculous math that grad students in physics lose sleep over. It's a systematic process, but not a neat and simple one.

To save you some time and tuition money, I'll just give you the quick rundown. Here's a rough description of what you'll see if you scatter some particles across space and run the simulation.

In the first instant, there's a huge burst of activity. Most of the seventeen types of particles are unstable, and they almost immediately undergo decay interactions, splitting into smaller, stable particles. After this initial explosion, you're left with only a few different types of particles, and only three you really need to keep an eye on: up quarks, down quarks, and electrons.

Next, as time creeps forward, patterns emerge. You start to see the quarks clump together in threes. There's no law of the Standard Model that says quarks have to clump in threes, but that's what happens. Trios of quarks interact with each other in a way that keeps them huddled up like that. As in the Game of Life, stable structures start to form over time by repeated applications of the same base rules.

In fact, this tendency toward trios is so strong that after the early excitement dies down, you pretty much never see a quark alone. Always in threes. Sometimes they'll be clustered in sixes or nines, or any multiple of three, but most of the time it's just three, flying together in a straight line. At this point, the word "quark" isn't that useful anymore. You're better off talking about the three-quark chunk as a whole, to save breath. So we coin some new terms. Two up quarks and one down quark, that's a "proton." Two down and one up? "Neutron."

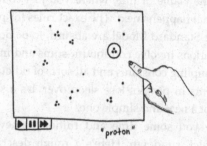

"*proton*"

Then what happens? More patterns and regularities emerge from the interaction rules.

As you watch, you notice positive charges and negative charges drifting together, while like charges drift apart. Again, this isn't in the rules. A particle over here doesn't "know" the charge of a particle over there. They just interact with other particles in their local surroundings, getting rerouted as they do. And these reroutings have a bias that adds up over time: Positives move gradually toward negatives and away from other positives.

This is happening very slowly, though, so let's speed up the simulation. Now this slow drift looks like a sharp tug. You see an electron (negative) falling quickly toward a "proton" (net positive). It speeds up as it gets closer, so fast that it actually speeds right past the proton. As it zooms away, it's slowed down, pulled backward now by the proton, until it changes direction altogether and does another flyby. And on and on, buzzing back and forth around the pull of the proton.*

You see this happen all over the space, anytime a proton and electron find each other. It's such a common structure, you might as well give it a shorter name than "electron buzzing back and forth around a proton." Some kind of name like "hydrogen."

And sometimes, remember, it's a bigger clump of quarks, a six or a nine or more. These cases are rare, but they do happen, and these bigger clusters pull even more electrons into their orbit. We can give each of these little systems a name, depending on the total amount of charge in the clump, names like "oxygen" and "chlorine" and "gold."

Maybe you already know what happens next. Speed up the clock even more, until the electrons are a blur, and you'll see these entire systems (let's call them "atoms") are drifting slowly through space. Sometimes they drift past each other unbothered, but sometimes they stick together and start drifting as a unit. You'll notice that hydrogen and hydrogen love to drift together as a pair, while oxygen prefers to drift with a hydrogen stuck to either side.

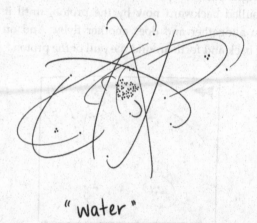

" water "

We still haven't introduced any new rules. This is the same simulation playing, over longer and longer amounts of time. Every time we observe some "new" phenomenon, it's always still explicable in terms of the base rules. Bonding, for instance? That's electrons following their interaction rules. When two hydrogens are close together, their electrons naturally start orbiting both protons, holding them together. Only when you zoom out and speed up does it look like a new rule: "Hydrogen travels in pairs."

You probably see where this is going, so let me get straight to it. These new mega-structures—"molecules," let's say—also behave in certain predictable ways, and they sometimes form gigantic mega-molecules: fats, proteins, lipids, ribonucleic acids, a whole zoo. And these each have their own characteristics and behaviors, and they sometimes form even bigger structures called organelles, which combine to form even bigger structures called cells. (Zoom out, speed up.) Some cells are off by themselves, while others interact with each other in units called organs, which in turn interact with each other in units called organisms. Some organisms cluster together in social groups or institutions, which cluster together to form classes or tribes, which interact to form an entire society. When societies interact, that's called history—and that's about as far as I can take this story.

Because it is a story, isn't it? Clearly I'm overplaying my hand here. No one's ever run a physics simulation that actually generated human societies, or even basic cell structures. How could they? It's not possible. The number of things to keep data on is literally the number of particles in the universe, so there's simply not enough room.

It's a story, yes, but it might be a true story! At the very least, it's a story with a lot of true elements to it. Each step in this chain is inferred from some very successful scientific model. Chemists believe that water is made of two hydrogens and an oxygen, and that theory hasn't made a wrong prediction yet. Behavioral economists believe you can explain people's economic behavior in terms of psychological and neurological factors. It's like a long relay race, where each field of study takes over for a lap.

= $\oplus$

= C

$$H_2C=CH, \quad HC, \quad C-H, \quad C=C, \quad H, \quad H = \hexagon$$

Still, it's entirely reasonable to believe this isn't the whole story. There are gaps in our understanding, which you may find suspicious. No one can really say they know *exactly* how human behavior arises from electric flashes in neuron circuitry. Artificial intelligence makes the idea plausible, but we haven't worked out the precise mechanics. You can take this as an opening to argue that there's something else going on here, some secret sauce that gets added at the level of human brains, which can't be explained in terms of the interactions of quarks and electrons.

Most math-oriented people I've talked to, though, seem to be broadly of the mind that something very close to this story is true. They believe the gaps are incidental and will eventually be filled in. So much has already been explained in terms of simple mathematical models: the motions of the stars, the diversity of life on Earth, natural disasters and the weather, the formation of the entire solar system. Why should we think the rest is any different?

Philosophers call this worldview "naturalism" or "scientific naturalism" and it's worth thinking about the implications. If this is true, if scientific naturalism is correct, then all of reality obeys strict mathematical rules. The entire universe must be identical to some carefully calibrated automaton. Everything going on around you, not to mention inside you, is a direct mathematical consequence of the laws of nature plus the initial configuration of the universe.

Which is a pretty trippy thought.

It raises some big philosophical questions, to say the least. If you buy into some version of this naturalist framework, here are three things to wonder about on your next commute.

Are these mathematical rules actual, bona fide Laws of Nature, somehow governing the progression of the universe? Or does the universe exist and change in time as a brute fact, and these "rules" are merely patterns we've found in it?

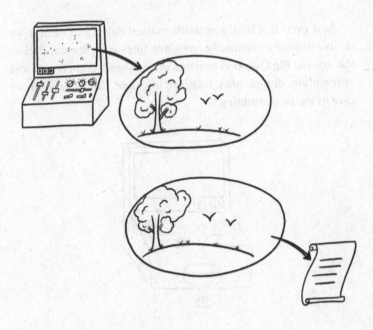

In either case, why *these* rules? They seem so bizarre and arbitrary. Why should this universe exist and not some other one? Does every conceivable mathematical universe exist in the same way this one does? Or are we somehow special, uniquely selected among the possible worlds to be instantiated, concrete, real?

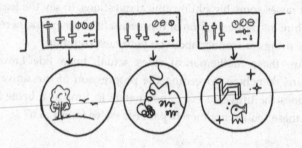

And even if it is all just mathematical rules, and even if we do live in what's essentially one giant, ultra-complex simulation, the age-old Big Question remains unanswered. Is there any sort of intention, design, plan, intellect, foresight, desire, warmth, or care in the programming?

I doubt we'll find answers to these questions anytime soon, and they may not even have "answers" in the standard sense of the word. At the end of the day, all we have are models we've invented, and each model is limited to a certain scope.

This model, the Standard Model, certainly aims higher than music theory or economics. It makes numerical predictions with a precision exceeding ten decimal places, which repeatedly come true in experiments. It offers a unifying explanation for nearly all phenomena observed in nature, collating and deepening the various pictures given by other models. And it comes bundled with a comprehensive story, this vision of reality as the smacking together of zillions of tiny dots, which many people find beautiful, humbling, even awe-inspiring.

But it's not everything. It has blind spots. I mean, the current Standard Model doesn't even explain gravity! (String theorists are working hard to fix this embarrassing oversight.)

Maybe it's not surprising that we should be able to find a mathematical object that so closely mirrors our reality. The ultimate aim of theoretical math is to collect and analyze all possible models, all possible structures and shapes and systems, all forms of logic and argument, under the same roof. It attempts to translate every conceivable and unconceivable *thing* into a common language, one universal set of notations and techniques. It's a project that on its face seems outrageous and impossible. Its continued success in explaining and predicting the phenomena of our daily lives is a curious blessing we don't fully understand.

At the very least, it's an interesting thing to think about.

Technically . . .

Pg. 23: We need to distinguish between compact and non-compact manifolds. This is only the complete list of compact sheet-manifolds—except the plane, which is non-compact. Additional non-compact manifolds include infinite manifolds, like the infinite cylinder; manifolds with invisible "open" boundaries, like a disk with its outer circle deleted; and some other weird creatures like an infinite-holed torus of finite size.

Pg. 59: The two end-points of the finite-length continuum aren't matched with anything, so this isn't actually a perfect matching. Instead it shows that the finite continuum is at least as big as the infinite continuum. But the infinite continuum is clearly at least as big as the finite one too, so they must be the same size.

Pg. 66: There's a slight issue with this naming system. LRRRRR . . . and RLLLLL . . . both refer to the same (middle) point on the continuum. In fact, any point on a perfect half mark, quarter mark, eighth mark, etc., will have two names. So we don't actually know that the number of points on the continuum is the same as the number of LR-addresses—it could be less.

(cont.)

To prove that there are at least as many points as LR-addresses, consider a different naming system. Split in thirds each time instead of halves, using L for left, M for middle, and R for right. Each LR-address still translates to a point under this new naming system, and there are no overlaps this time. (E.g., the alternate name for LRRRRR . . . under the new system is MLLLLL . . . , which isn't an LR-address.) So there are at least as many points as LR-addresses.

Pg. 78: Any container that's the same shape (topologically speaking) as a sphere. A donut-shaped container, for instance, can contain a flow with no fixed point. This theorem is true in every dimension.

Pg. 108: It also can't have any "loops"—this proof only works if there's no way for the game to loop back and forth between the same positions indefinitely. Many games have "draw by repetition" rules, in which case the theorem still applies.

Pg. 124: To actually prove facts about primes, though, you need to add some axioms that define what exactly a "prime" is. These are just the five basic axioms, and each new concept you want to use will require more axioms.

(cont.)

Pg. 174: The length of a day is not exactly simple harmonic motion, but it's a very close approximation. There's a small error term which becomes more significant as you get further from the equator. In the Arctic and Antarctic, the approximation breaks down completely and the sun circles the horizon for months at a time.

Pg. 193: I'm leaving out a key element of the model here. If these were actually the rules, the electron would gradually lose energy and fall into the nucleus. In the actual Standard Model there's a minimum "quantum" of energy that stops this from happening.

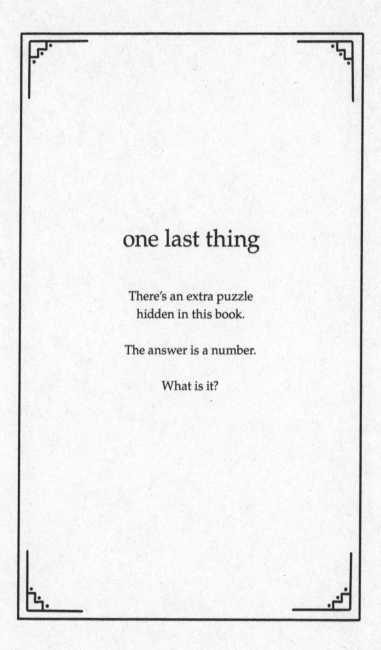

one last thing

There's an extra puzzle
hidden in this book.

The answer is a number.

What is it?

About the Author

Milo Beckman has been addicted to math since a young age. Born in Manhattan in 1995, he began taking math classes at Stuyvesant High School at age eight and was captain of the New York City Math Team by age thirteen. His diverse projects and independent research have been featured in *The New York Times*, *FiveThirtyEight*, *Good Morning America*, *Salon*, *The Huffington Post*, *The Chronicle of Higher Education*, *Business Insider*, *The Boston Globe*, *Gothamist*, *The Economist*, and others. He worked for three tech companies, two banks, and a US senator before retiring at age nineteen to teach math in New York, China, and Brazil, and to work on this book.

About the Illustrator

The role of the artist is to make the revolution irresistible.

—Toni Cade Bambara

M is a brown genderqueer cultural worker and organizer. Under the name Emulsify, they create art that helps them heal, learn, advocate, and imagine new worlds. They believe *all* art is powerful and political. M lives in Brooklyn with their wife and spends a lot of time creating while snuggling their pups. M's creative energy and love are a part of all they do: they are a trained abortion doula, founder of Emulsify Design, and creative director of Arrebato, a space for Queer Trans Black & Brown community. Through their work, M has made incredible friendships, learned from brilliant peers, and found their home. You can follow M's work on their website at emulsify.art.